Wireless Recon...........
in Penetration Testing

Wireless Reconnaissance in Penetration Testing

Matthew Neely

Alex Hamerstone

Chris Sanyk

ELSEVIER

AMSTERDAM • BOSTON • HEIDELBERG • LONDON
NEW YORK • OXFORD • PARIS • SAN DIEGO
SAN FRANCISCO • SINGAPORE • SYDNEY • TOKYO

Syngress is an Imprint of Elsevier

Acquiring Editor:	Chris Katsaropoulos
Development Editor:	Meagan White
Project Manager:	Mohanambal Natarajan
Designer:	Russell Purdy

Syngress is an imprint of Elsevier
225 Wyman Street, Waltham, MA 02451, USA

Library of Congress Cataloging-in-Publication Data
Application submitted

British Library Cataloguing-in-Publication Data
A catalogue record for this book is available from the British Library.

For information on all Syngress publications
visit our website at www.syngress.com

ISBN: 978-1-59749-731-2

Printed and bound by CPI Group (UK) Ltd, Croydon, CR0 4YY

Transferred to digital print 2012

Dedication

I'd like to start out by thanking Joan Amaratti for believing I could write a book all those years ago. I'd also like to thank Ken Stasiak and the SecureState family for supporting me throughout the entire writing process. Finally I dedicate this book to Meagan Call for being a wonderfully supportive wife through this and all my projects.

--Matt

I dedicate this book to BNH, ELH, and JAH.

--Alex

Contents

Author Biography

Matt Neely (CISSP and CTGA) is the Director of Research, Innovation and Strategic Initiatives at SecureState, a security management consulting firm. At SecureState Matt leads the Research and Innovation team which focuses on imagining, researching and developing new offensive and defensive capabilities. His research interests include the convergence of physical and logical security, lock and lock picking, cryptography and all things wireless.

Mr. Neely is actively involved in public speaking and has spoken as a subject matter expert over seventy-five times at various local, national and international conventions and user group meetings including BlackHat EU, DefCon, ShmooCon, Thotcon and Notacon. Mr. Neely also guest lectures at local colleges on topics on security and risk management. He is a founding member of the Cleveland Chapter of TOOOL and is a host on the Security Justice podcast.

Alex Hamerstone is the Compliance Officer for TOA Technologies, an international workforce management software company. He is an RABQSA certified ISO27001 Auditor and is active in the security community. When he isn't working or writing, he enjoys tinkering with electronics and spending time with his family.

Chris Sanyk is an IT professional with over twelve years of experience in everything from desktop publishing and web design, PC and server hardware, to user support, system administration, and software development. In his spare time, he blogs and develops video games at his website, csanyk.com.

Preface

Radio waves surround us and more and more devices are being made wireless. Most penetration testers focus only on the very small portion of the radio spectrum using by 802.11 and Bluetooth devices. Physical penetration tests often miss guard radios, wireless headsets, wireless cameras, and many other radio devices commonly used in the modern corporation. These systems transmit a wealth of information which can aid a penetration tester in a targeted attack.

This book aims to educate penetration testers on how to find these too often ignored radios and mine them for information. The following chapters include information ranging from choosing the best equipment to use and how to find frequency information, to actual case studies demonstrating how this information has been used during penetration tests. The authors draw on a combined knowledge derived from performing hundreds of penetration tests and decades of radio experience to share tips, tricks and helpful notes about this less explored avenue of attack. This book is the definitive resource for anyone interested in adding radio profiling to his or her arsenal of penetration testing tools.

The book is also a great resource for the people who need to defend computer systems and companies. Like penetration testers, defenders often ignore wireless traffic outside of 802.11. This book shows various radios that might be deployed in various environments and how attackers could exploit the information leaked by these radio systems. Essential information on how to prevent this information leakage from occurring is also included.

HOW THIS BOOK IS ORGANIZED

The best way to read this book is in the order it's presented, but the chapters are structured to assist the reader should he decide to read out of order. When key concepts are mentioned which were covered in earlier chapters, the page will reference the earlier chapter so the reader can flip back if he needs a refresher or is reading the chapters out of order. A glossary is also included at the end of the book to help with unfamiliar terms.

Chapter 1: Why Radio Profiling?

In the first chapter of the book the reader will learn what radio reconnaissance is and how it is useful during penetration tests. The chapter concludes with a short case study of radio reconnaissance used during a physical penetration test at a power company.

Chapter 2: Basic Radio Theory and Introduction to Radio Systems

In Chapter 2 the reader will learn the theory behind how radios work and gain an introduction to the different radios systems you will encounter while performing radio reconnaissance. The chapter starts by discussing basic radio theory. The authors cover the terminology needed to understand underlying concepts, give an overview of the radio spectrum and discuss how radio waves behave at different frequencies. Next, they cover how a radio works and the different components found in a radio receiver. After the reader learns how radios work, she will read about the most important part of a radio: the antenna. This part of the chapter starts out by covering antenna theory and wraps up with a discussion of the most common types of antennas one might encounter while performing radio reconnaissance during a penetration test. After antennas are discussed, the chapter moves on to the different ways radios encode data (modulation types). This section covers analog, digital and spread spectrum modulation types. Next is a rundown of the different types of most commonly used radio systems. This starts out simply, discussing simplex verses duplex radio systems, expands to cover repeaters, and concludes with an explanation of trunked radio systems. The chapter ends with recommendations on where to learn more about radios and radio theory.

Chapter 3: Targets

In Chapter 3 the reader will learn about some of the different types of targets which could be searched for during radio reconnaissance. Highlighted targets include guard radios, cordless phones and video cameras.

Chapter 4: Offsite Profiling

The Offsite Profiling chapter covers how to gather as much information as possible on a client's radio systems before arriving onsite. The authors suggest terms to use in online searches, how to search the FCC license database, and specialty websites that can be used to gather more information on the client's equipment. The chapter concludes with a case study covering the offsite profiling performed before a physical penetration test of a ship dock and how the information was used during the attack.

Chapter 5: On Site Radio Profiling

Chapter 5 continues on to the next step on a penetration test and covers information that can be gathered on a target's radio systems onsite. The chapter starts out with radio related items to keep an eye open for while onsite and what a penetration tester can learn from these items. Next the chapter explores frequency counters and how to use one while profiling a target. Next the reader learns what can be discovered just by looking at a targets radio systems and antennas. The authors discuss what can be learned about the make, model and type of radio used and how to estimate the frequency range, based off of the radio's antenna. Finally, this chapter also includes common frequencies and frequency ranges to search while onsite. The chapter concludes with a case study of intercepting wireless headsets used at an insurance company and the information this provided about the company's internal network.

Chapter 6: How to Use the Information You Gather

In this chapter the reader will learn how to use the information gathered monitoring the targets radio systems. It includes specific advice on using the information gathered from guard radios, wireless headsets and phones and wireless cameras.

Chapter 7: Basic Overview of Equipment and How it Works

Chapter 2 explored the scanner and how it works. In Chapter 7 the authors cover the common controls and features found on scanners and how to operate them. Next the book explains how to select a scanner for wireless reconnaissance and provides recommendations on scanners to use for wireless reconnaissance. After recommending specific radios the authors also discuss and recommend the antennas which they have found the most valuable over the years. The authors conclude this chapter by discussing accessories a reader may want to add to his wireless kit for radio reconnaissance.

Chapter 8: The House Doesn't Always Win: A Wireless Reconnaissance Case Study

Chapter 8 is a case study that pulls together all the information provided in the book. During this case study the authors outline how wireless reconnaissance was invaluable during a physical penetration test on a casino. They start out by showing offsite profiling using the techniques discussed in Chapter 4. Next they tell how they used the offsite information and then expanded it using the onsite profiling techniques discussed in Chapter 5. Throughout this case study the reader will see firsthand how the information gathered by monitoring the casino's radio systems was key to the successful penetration of this high value target.

Chapter 9: New Technology

During Chapter 9 the reader will learn where the world of wireless reconnaissance and penetration testing are heading. The authors discuss the shift to digital transmissions, the challenge this presents to penetration testers and some ways to overcome these challenges. Next they talk about Software Defined Radios and how this technology will bring about a new golden age of wireless hacking. In this section the reader will learn what a software defined radio is, common commercial and open source software defined radios and some examples on how they can be used. Next the book covers the trend of VOIP enabling radio dispatch systems. This section includes a case study showing the security problems seen by the authors in VOIP enabled dispatch systems. The chapter ends with recommendations on resources to keep up to date on scanners, wireless reconnaissance, and wireless security.

CHAPTER 1

Why Radio Profiling?

Information is everywhere, if you know where to look. When performing penetration tests, uncovering the correct information during the reconnaissance phase can often mean the difference between a successful test and failure. While many of us are familiar with the often used data gathering methods employed by penetration testers, radio traffic can provide a great deal of valuable information. This rarely used reconnaissance method, when used effectively, can provide a wealth of data. The information gathered by the methods described in this book is useful for both physical and logical penetration tests.

In addition, as with any other methods used by penetration testers, understanding the methods that can be used by penetration testers and attackers is useful when securing networks and facilities. To protect against attackers, it is necessary to think like an attacker.

Not everything in this book will work in every situation, which is of course not unique of this method of reconnaissance. However, as the included case

CONTENTS

NOTE

This book assumes that you are familiar with the basic concepts of penetration testing. Physical penetration testing is the process of testing the physical security of an organization or facility, while logical penetration testing is the process of testing the network and computer security of an organization or facility. Often, physical and logical penetration tests are combined; for example, once a facility is penetrated, we will then use the physical access to plug into the network or physically access computing equipment.

studies will show, when the methods in this book are used the results can be immensely valuable.

The equipment necessary to perform what is described in this book doesn't have to be expensive. While there are radios costing thousands of dollars, a basic receiver purchased second hand can provide much of the functionality that you will need. Once the basics are mastered, a determination can be made as to whether to invest in more expensive and more complex equipment. Where possible, multiple methods using varied equipment will be described, with a focus on practicality.

Penetration testers and attackers tend to spend a lot of time looking at 802.11 and other wireless networks, and occasionally will look for Bluetooth to see if there is any valuable traffic on devices such as keyboards. This is only the beginning when it comes to what is available on the radio spectrum. Figure 1.1 shows the radio spectrum (3 kHz–300 GHz) as it is divided up in the US and highlights the portions of spectrum used by 802.11 and Bluetooth. As you can see, these services use just a fraction of the entire radio spectrum. Figure 1.2 shows the radio spectrum, as well as the radios and wireless devices that most penetration testers miss.

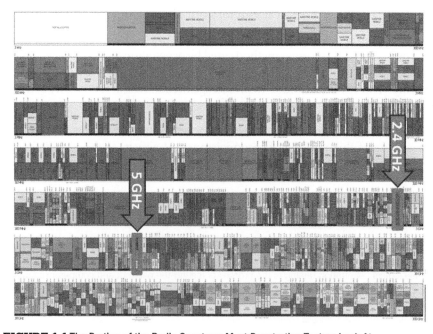

FIGURE 1.1 The Portion of the Radio Spectrum Most Penetration Testers Look At

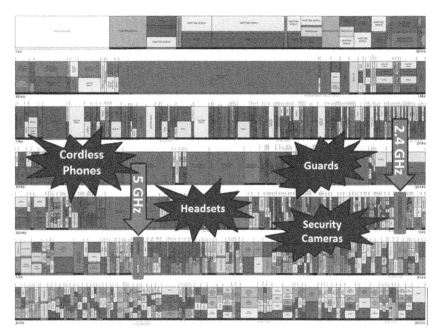

FIGURE 1.2 What Most Penetration Testers Miss

GUARD RADIOS, WIRELESS HEADSETS, CORDLESS PHONES, WIRELESS CAMERAS, BUILDING CONTROL SYSTEMS

The targets on the radio spectrum consist of those that have been around for decades, such as the two-way radios used by guards, and those that are just beginning to proliferate such as wireless video cameras. Some of the target radio traffic may have an obvious use for a security professional, such as Bluetooth keyboards. The ability to capture keystrokes can be invaluable for clear reasons. Other traffic, however, may have less obvious advantages. Later chapters will cover the details of on- and off-site reconnaissance, and how to use the appropriate equipment. It is, however, important to first gain a basic understanding of the types of information available to an enterprising attacker. If the target organization has a guard force, the guard's radio transmissions provide a wealth of intelligence. From the guard's names, to the time of shift changes, to internal jargon, there is much to glean. When launching a social engineering assessment, or attack, knowing the guard's names adds credibility to the penetration tester or attacker. Listening in to guard traffic may also let the attacker know when the guards will not be at their posts, either because of scheduled rounds or unscheduled bathroom or smoke breaks. To take things further, in combination with a police scanner, an attacker can learn the response times to incidents. Knowing the time between the discovery of an incident and alerting of authorities, and then authorities' response time can let an attacker know how long they can be inside the facility without being caught.

Traffic from wireless cameras can provide much of the same information as traffic from guard force radios. Knowing where the guards are within the facility or grounds, and which areas are unoccupied can mean the difference between success and failure during a physical penetration assessment. Additionally being able to see the inside layout of a building before you step inside of it can also be invaluable when performing a physical penetration test. While far less likely to occur in the real world than in Hollywood, it may also be possible, depending on camera resolution and angles, to be able to view cipher lock codes from the camera transmission.

In addition to profiling and reconnaissance, this book also offers valuable insight into counterintelligence. Understanding what information leaks unintentionally from your organization will help to ensure that confidential information remains confidential. The authors have been involved in situations where confidentiality was essential, and have discovered information in unlikely places. One example was while sweeping a conference room for bugs wireless microphones were discovered. The conference room was to be used for a presentation about a potential corporate merger. Despite a large security budget and bug sweeping teams, had wireless microphones been used during this high level meeting, anyone within the vicinity would have been able to listen in on the entire presentation.

Before trying anything in this book, make sure that you understand the legal and ethical ramifications of your actions. There are certain things that are always illegal, such as interfering with radio transmissions, and there are many other things that are illegal in most circumstances. Be sure to seek legal council prior to getting in too deep. Of course, as security practitioners, it is often frustrating

TIP

It is extremely important that when performing any type of penetration assessment the scope and ground rules are agreed upon in writing prior to starting. Be sure to stick to the scope. While you may find additional items of interest while profiling, only assess those that are within scope. Consider getting a "get out of jail free" card or letter from the organization that you are assessing. If security or law enforcement catches you, the letter can be presented to explain that you are a security professional on a contracted engagement, and not a common criminal. Include the names, titles, and contact information of at least three people at the organization who know that you are performing an assessment. Also, be sure to let those individuals know to keep their phones nearby and to answer them no matter what the time.

that we are bound by the law while attackers, by their very nature, are not. This means that it isn't possible to attempt everything an attacker would while staying within the law. Thus, it is essential to understand the illegal tools and techniques that attackers have at their disposal to understand how to defend against them.

CASE STUDY

Perhaps the best way to understand the true value of radio reconnaissance is with a case study. While this case study includes a fictionalized version of events, the authors on actual engagements have used successfully all the techniques described in the following paragraphs.

We knew we were lucky that the power company's fence was not electrified. Bad jokes aside, when attempting to enter a fenced power company facility, the tools that come to mind may be bolt cutters and carpet to throw over barbed wire rendering it useless. In this case, we had those with us, but it turned out our radios would also prove valuable. The first thought when you hear information security is probably not a couple guys dressed in black tactical gear in the woods up to their ankles in cold mud. In today's global economy, the stakes are high and competitors and criminals will often stop at nothing to gain the upper hand or steal and sabotage information and equipment. As networks become hardened and information more protected, many attacks have moved to the physical realm. It is often cheaper for nefarious corporations or overseas criminals to send operatives to facilities and attempt to steal information than it is to hack through the network. The goal of a penetration test is to find vulnerabilities and help to mitigate them before attackers can take advantage. On this dark night, that put us in the woods.

The irony is that our target was an energy company, a fact not lost on us as we shivered in the cold. The main gate was guarded, so we followed the fence through the woods, and waded through a cold creek. Our reward was discovering a break in the fence. The scraps of carpet we had in our bags remained there. It is an old trick, but a well-known one, that placing a scrap of carpet over barbed wire makes scaling the fence a breeze. Twenty yards away, a small building stood alone in a field on the property. The door swung open with just a twist of the handle—it was not locked. Inside we found a few company shirts, and a breaker panel. We left the panel alone, because we are the good guys, and didn't know what dangers we could cause by flipping off the main switch. We discovered later that it controlled all the parking lot and perimeter lights—very useful for a malicious attacker. We moved toward the main facility unimpeded, and reached a locked door. The lock was one we knew well—not this particular lock of course, but the make and model. It took us under a minute to pick it and gain access to the building. Breaching the perimeter and gaining access to the main facility was our proof of concept and ended this portion of the assessment. But while we were shivering outside, a second team was taking another avenue to compromise the company, and what they found once they were inside let us know what we could have done inside the building.

People are naturally trusting. And they usually want to help. The second team drove right up to the guard gate, and used a technique more effective than ramming the fence at full speed: conversation. The team told the guard that they were there to fix a network issue. The guard asked their names and what company they were with. We gave them our real names, and made up a company name. The guard diligently printed our names and made up company on visitor badges, and was even kind enough to direct us to the building that houses the computer room. Exiting the car, the team then walked right through an unlocked door, up a few stairs, and found another unlocked door. Behind this door was what we had come for, the server room. All the equipments were neatly and carefully labeled by the company's IT staff. The backbone of the network. The life safety systems. The financial accounting systems. We added an administrative account to a few machines using a password and username that was on the bulletin board, pinned right next to a poster advertising the importance of maintaining confidentiality. While this case study may not seem to have much to do with radio, aside from the fact that the teams used them to communicate, radio turned out to be the difference between success and failure on this penetration test. Before coming onsite we found the frequency used by the company guard force by using a search engine and scouring radio hobbyist Web sites. While this may seem like a small detail, it allowed us to monitor the guard communications, and slip offsite as soon as they realized something was amiss. Had we not been monitoring the communications, we would likely have been caught and the penetration test would be considered a failure by the client. Instead, we were able to add another successful breach to our history. In this book you'll learn the tools and techniques we used to discover the frequencies the guards used and monitor them.

CHAPTER 2

Basic Radio Theory and Introduction to Radio Systems

Whether you credit Guglielmo Marconi or Nicola Tesla with its invention, radio has been in use for over a century, and will continue to grow for the foreseeable future. Indeed, tried and true radio technologies are becoming even more important in our increasingly connected world.

A solid background in radio terminology and theory is essential for wireless reconnaissance. This chapter will provide a general overview of the fundamental concepts of the science and technology.

THE ELECTROMAGNETIC SPECTRUM

In the nineteenth century, electricity and magnetism were understood to be two separate phenomena, until discoveries by Michael Faraday, James Maxwell, Heinrich Hertz, and others unified these two forces under a single theory of electromagnetism. Electromagnetism is concerned with the forces that occur between electrically charged particles, and today is considered one of the four fundamental forces by modern physics. This paved the way for the inventions of Marconi and Tesla only a few years later, and an accelerating avalanche of innovation that continues to this day. In the twentieth century, a continuing trend of reductionism in the field of physics still endeavors to unify theories of the four fundamental forces (gravity, electromagnetism, weak, and strong nuclear forces) into one Grand Unified Theory. However, all technological applications of radio that are of interest to security reconnaissance do not depend on such advanced physics, and can be well understood using the groundwork provided by Maxwell and his contemporaries.

CONTENTS

Terminology

It's important to have a firm understanding of the underlying science and the terminology used to describe it. The following concepts are fundamental to understanding radio, and it is critical that you become familiar with them.

1. *Frequency:* Frequency is the measure of how many times the radio wave oscillates in a unit of time. Looking at the graph of a sinusoidal wave form (Figure 2.1), frequency can be understood as the time between like portions of a wave (such as the peak or the trough) as the wave passes over a stationary point in space.
   Frequency is measured in Hertz (Hz). It is a measurement of frequency, regardless of the medium—Hertz is used not only for measuring the frequency of electromagnetic waves, but in other contexts as well, such as acoustic waves and seismic waves. In every case, 1 Hz = 1 cycle per second. In the electromagnetic spectrum, the band known as radio waves ranges from about 3 kHz to about 300,000 MHz. Named after German physicist Heinrich Hertz, Hertz was established as a term in 1930, officially adopted as an SI unit in 1960, and widely replaced the phrase "cycles per second" by 1970. When an SI unit is spelled out in English, it should always begin with a lower case letter (hertz), except where any word would be capitalized, such as at the beginning of a sentence or in capitalized material such as a title. As a part of the SI system of measurement, the typical prefixes kilo—(1000), mega—(1,000,000), giga—(1,000,000,000, or 1 billion) are used in the customary fashion to create the derived units kilohertz (kHz), megahertz (MHz), and gigahertz (GHz). The first two should be familiar to anyone who has seen an AM/FM radio; gigahertz has become familiar in the last decade, as 802.11 (Wi-Fi) operates in the 2.4 and 5 GHz band.

2. *Wavelength:* Wavelength is the linear distance between two like parts of the wave form, typically the peak or trough of the wave (Figure 2.1). Because all electromagnetic waves travel at the speed of light, an inverse relationship exists between wavelength and frequency. The longer the wavelength, the lower the frequency, and vice versa. This is because a wave of a higher frequency moves through its cycle in less time (thereby creating more cycles per second), and all electromagnetic waves travel through a vacuum at the same speed, but with a high-frequency wave completing a single oscillation in less time, it necessarily covers a shorter distance, resulting in a shorter wavelength.

3. *Amplitude:* Amplitude is a measure of the energy in the wave (Figure 2.1). Amplitude is directly related to the strength of the signal and the amount of energy in it. A strong signal has a high amplitude, while a weak signal has a low amplitude. When a signal is *amplified*, its amplitude is multiplied.

The radio spectrum is a vast band of the electromagnetic spectrum which includes frequencies from 3 kHz to 300 GHz. This is a huge chunk of the spectrum, and is commonly divided into different bands.

Common Frequency Ranges of the Radio Spectrum

Band	Frequency in Hertz (Hz)
Extremely low frequency (ELF)	Below 3 kilohertz (kHz)
Very low frequency (VLF)	3–30 kHz
Low frequency (LF)	30–300 kHz
Medium frequency (MF)	300–3000 kHz
High frequency (HF)	3–30 megahertz (MHz)
Very high frequency (VHF)	30–300 MHz
Ultrahigh Frequency (UHF)	300–3000 MHz
Superhigh frequency (SHF)	3–30 gigahertz (GHz)
Extremely high frequency (EHF)	30–300 GHz

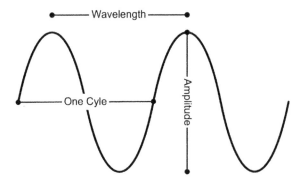

FIGURE 2.1 Sinusoidal Wave Form

PIPE ORGAN ANALOGY

It can be difficult to think about electromagnetic waves, since they cannot be seen or felt. Fortunately, wave phenomena behave very similarly in other media. For example, with sound waves, waves of physical matter (usually air) vibrating, we see the exact same relationships between frequency and wavelength. Consider the pipe organ: The vibrating pipe creates a vibrating column of air vibrating in sympathy with the pipe. The pipe has a natural frequency, based on its length, which is tuned to a specific note. When energy from the organ causes the pipe to vibrate, it tends to vibrate at its natural frequency. The low notes (low frequency) are sounded by very long pipes (long wavelength), while the high notes (high frequency) are sounded by very short pipes (short wavelength). If the organ is played at full volume, more energy is imparted into the pipe, resulting in greater physical distance traveled in each cycle of vibration, in other words higher amplitude, and a more energetic vibration. Keep this image of the pipe organ in mind (Figure 2.2); it will become important later on when we discuss antennas and tuning to resonate on a target frequency.

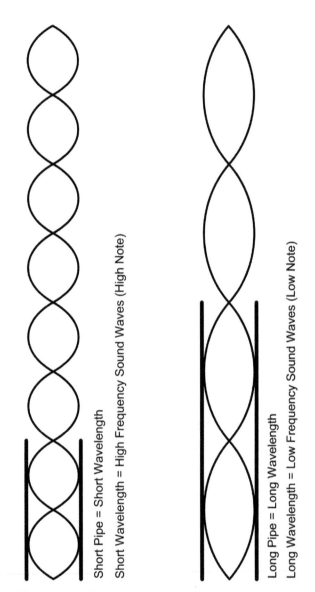

FIGURE 2.2 Different Length Pipes Generate Different Frequencies

NOTE

Regardless of the frequency, radio waves all travel at the speed of light, 186,280 miles per second (3.0×10^8 m/s)!

> **NOTE**
>
> Wavelength is equal to the speed of light divided by the frequency

The entire radio spectrum is vast, a frequency band spanning almost 300,000 MHz. By comparison, the FM band, the one used in the US for commercial radio broadcast, is only a 20 MHz wide portion of that 300,000 MHz range. The RF portion of the electromagnetic spectrum covers the lower-end frequencies. Above 300 GHz, electromagnetic waves shift into near-infrared, infrared, visible light, ultraviolet light, and finally, X-rays, and gamma rays, where frequencies get into the terahertz range and beyond. At ultraviolet frequencies and higher, electromagnetic radiation becomes increasingly dangerous due to the energies involved in generating these frequencies, and the ability of tiny wavelengths to penetrate most solid matter, and requires special shielding and careful handling. Of course, even lower frequency electromagnetic waves can pose a danger at higher intensities and under the right conditions. For example, 2.4 GHz waves are used in microwave ovens, infrared is used in heat lamps, and most lasers are in the visible spectrum.

Wavelength/Frequency Characteristics

From our quick overview of the electromagnetic spectrum, it's readily obvious that electromagnetic waves behave very differently at different frequencies and wavelengths. Even within the RF portion of the electromagnetic spectrum, radio waves behave differently depending on their frequency as well. Understanding how radio waves behave at different frequencies is helpful for understanding how the target signal is likely to react, based on the frequency, material in the area, and atmospheric conditions.

An important property of different frequency RF signals is the distances they are capable of travelling. Due to the different wavelengths of different frequencies of RF signal, they are absorbed and reflected differently by different types of material. The higher frequencies are "line-of-sight" frequencies, but tend to be absorbed and blocked by solid objects. Waves of RF frequencies in the shortwave band are of a length that causes them to bounce off the Earth's ionosphere (Figure 2.3). The lower-frequency waves actually bend to follow the curvature of the Earth. Each of these frequencies is useful due to these unique properties which give them capabilities that make them applicable in specialized contexts.

VHF and UHF frequencies travel line-of-sight, with a typical range of 20–30 miles. This distance varies greatly, depending on obstructions, as VHF and UHF are easily blocked by buildings or topography. Antenna height and transmitter

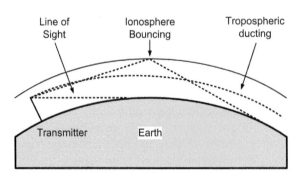

FIGURE 2.3 Behavior of Radio Waves of Different Frequencies

THE ATMOSPHERE

Atmospheric and meteorological conditions can have a significant effect on the propagation of radio waves. Earth's Atmosphere consists of several layers, each with its own properties. Each layer absorbs, reflects, and refracts electromagnetic waves differently, giving rise to a number of interesting phenomena which can aid or hinder radio operators. The principal layers of the atmosphere, starting from Earth's surface, are: the Troposphere, Stratosphere, Mesosphere, Thermosphere, and Exosphere (Figure 2.4). At frequencies of ultraviolet light and above, electromagnetic radiation possesses enough energy to dislodge electrons from atoms, creating ions. Above the stratosphere, and extending through the mesosphere and partially into the exosphere, is the ionosphere. Reaching from altitudes of 50–1000 km above sea level, this region is so named because solar radiation interacts with air molecules, exciting them and causing them to become ionized (electrically charged). These charged layers of air molecules interact with electromagnetic phenomenon, such as radio waves and Earth's magnetic field. The ionosphere consists of four layers (Figure 2.5):

F—F1 and F2 merge together at night.
E—Weakens at night.
D—Closest to the Earth. Disappears at night.

The amount of ionization and number of layers varies greatly depending on the radiation received from the sun. At night, when the Earth blocks the sun's radiation from reaching the dark side of the planet, the F1 and F2 regions merge together. The E and D regions also become weaker at night as the level of ionization decreases, and the D-layer disappears. This allows HF (below 30 MHz) waves to reach the F-layer, where it reflects due to the wavelengths of RF emissions at these frequencies. This is why shortwave radio can be heard from such distances. Another atmospheric phenomenon that results in altered behavior of radio transmissions is tropospheric ducting. This occurs when cold and warm air streams meet about 2 km, or approximately 1.25 miles above the Earth. This phenomenon, which is often seen during the summer and usually lasts about an hour at a time, creates a "pipe" of warm and cold air that reflects the signal repeatedly, in a zig-zag fashion, allowing the VHF and UHF RF signals to travel great distances. Tropospheric ducting of VHF frequencies start above 100 MHz. Below that, the signal quality is greatly deteriorated.

To summarize, RF signals below 30 MHz are capable of bouncing off of the atmosphere's upper layers and can travel great distances, reaching around the curvature of the Earth. This works best at night when the D-layer of the ionosphere disappears. Signals above 30 MHz are more line-of-sight, apart from the tropospheric ducting phenomena.

power are also key factors, which is why transmitters are often placed on the top of tall towers. The higher the antenna is placed, the longer distance to the horizon will be, following line-of-sight from the antenna.

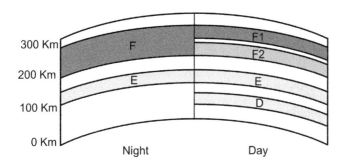

FIGURE 2.5 Ionosphere Layers

Exosphere	10,000 km
Thermosphere	690 km
	Aurora
Mesosphere	85 km
	Meteors Showers Visible
Stratosphere	50 km
	Weather Balloons
Troposphere	6 -20 km
	Airplanes

FIGURE 2.4
Atmospheric Layers

There are certain atmospheric phenomena that can change the behavior of radio waves. Sporadic-E, also known as E-Skip, is caused by the E-layer of the ionosphere becoming thicker (see Sidebar), which causes VHF waves to bounce off the ionosphere much like HF waves, and thus travel further than they would normally.

How Materials Affect Radio Waves

When performing radio reconnaissance, it is obviously essential to be able to receive radio signals. Therefore, it is helpful to understand how certain materials attenuate, or block radio waves. Where radio waves are concerned, matter can do one of two things: it can conduct radio waves and be a conductor, or insulate and be a dielectric. The majority of conductors are metals, while the majority of dielectrics are non-metallic. When a radio wave encounters a material, some quantity of its power will be reflected by the surface, some quantity of power will pass into the material, and some quantity will pass through the material. The amount of power absorbed by the dielectric is the material's *attenuation coefficient*. The quantity of energy that is able to pass through a dielectric is determined by the material's attenuation coefficient and thickness. A low attenuation coefficient will allow radio waves to easily pass through the material.

Multipath

Multipath occurs when the signal from a transmitter bounces around enough that it arrives at the receiver at different times (Figure 2.6). When this happens, signals arrive at different points in the phase, and interfere with each other. This is a common problem in urban environments where signals bounce off buildings and metal objects, and can cause deteriorated audio signals and "ghosting" in video images. One way to combat multipath is to have an antenna at the proper polarity. (See Antennas section of this chapter.)

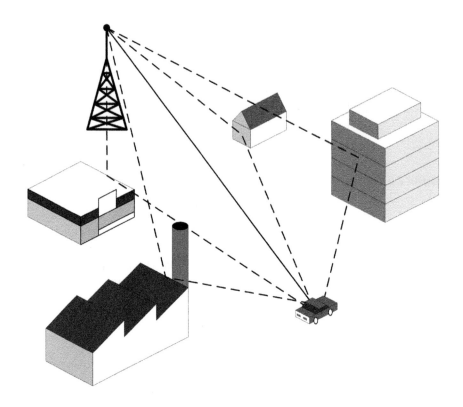

FIGURE 2.6 Multipath

With technologies such as 802.11 Wi-Fi (2.4 and 5 GHz), multipathing actually becomes a useful property. A Wi-Fi signal indoors will reflect off of surfaces (walls, cabinets, etc.) and reach corners which would otherwise be in shadow if a line-of-sight frequency were used. Wi-Fi radios have specialized circuitry which allows them to combine these multiple paths, shifting the out-of-phase wave forms arriving from multiple paths back together into a reconstituted signal that is stronger than what otherwise would have been received, allowing Wi-Fi to work at longer ranges than it otherwise would. Wi-Fi signals are partially attenuated by most building materials, and completely blocked by some, so this is an important factor in 2.4 GHz's usefulness in 802.11 applications.

REGULATORY AGENCIES

For wireless reconnaissance, we are mainly interested in a few chunks of the radio spectrum, specifically in the 30 MHz to 1 GHz range, 2.4 and 5 GHz spectrum. In the United States, the RF spectrum is governed by the Federal Communications Commission (FCC), who designates and apportions the

RF spectrum into different bands, which are commonly used by different types of services. The FCC also is responsible for coordinating and issuing radio operation licenses, which authorize organizations to transmit on specific frequencies, at specific power levels, in specific geographical regions. The FCC manages the RF spectrum, and therefore is required to keep and maintain public records of who is licensed to use which frequencies. This makes them a valuable resource for profiling targets. How to access and use this license information from the FCC will be discussed in Chapter 4. Additionally every electronic circuit emits an RF field when powered. Electronics manufacturers are required to certify products with the FCC, to guarantee that they are properly shielded so that their incidental RF emissions are properly contained and will not interfere with the operation of legitimately licensed broadcasts.

Other countries have similar organizations to the FCC which manage and license access to the spectrum in that country. In Europe, each country has its own governing body that manages the use of the spectrum within their borders. The ITU (International Telecommunication Union) is a UN agency that coordinates shared global spectrum. This body covers spectrum utilization over international waters, satellite, short wave amateur radio bands, and the broadcast shortwave spectrum. The ITU provides standards to help the various country regulatory bodies coordinate.

APPLYING THE SCIENCE: RADIO TECHNOLOGY BASICS

Having provided a basic understanding of the radio spectrum and how radio waves at different frequencies act, we can apply this knowledge to understand how information can be transmitted and received using radio frequencies. For this we'll learn the basic components of a radio and some basic radio theory.

While we all have an image when we hear the term radio—be it a clock radio, car stereo, or walkie-talkie—all radios do three basic things:

- *Tuning:* A radio can tune into a desired frequency.
- *Amplification:* A radio can amplify the weak signal that is received as the radio waves pass over the antenna. The amount of energy imparted when a radio wave passes over an antenna is very weak. So the signal usually needs to be amplified to a level where the other parts of the radio can process it.
- *Demodulation:* A radio must detect the signal over the background noise, and demodulate the signal into a usable form. In most cases, this means converting the signal into sound waves the end user can hear.

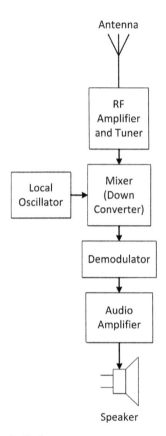

FIGURE 2.7 Block Diagram of a Radio

To accomplish these things, a radio has different specialized parts. Figure 2.7 is a block diagram for a Superheterodyne Receiver (also called Superhet), which is the most popular type of receiver used today, and commonly used in modern day scanners.

It is important to understand at least at a high level what is going on inside the radio, in order to understand the limits of your equipment, as well as troubleshoot issues, such as front-end overload and spurious or false signals.

Following the path an incoming signal from the air takes as it is processed by the radio receiver:

1. *Antenna:* When radio waves pass over the antenna they induce a small current into the antenna which passed into the radio. When an electrical conductor passes through an RF field, the field induces a current in the

conductor. Antennas and how they work will be covered in more depth later in the chapter.

2. *RF amp and Tuner:* The RF amp amplifies the weak signal that comes from the antenna. Often this signal is only a few microvolts. Getting the amplification right is a delicate balance for the engineer building the system. A strong enough amplifier is needed in order that the radio may be able to detect and demodulate a weak signal. But too strong an amplifier will create a signal the overloads the mixer, resulting in distortion in the signal as a variety of garbage signals are introduced into the radio and the radio generates signals outside of the intended frequency. This stage is also where the signal from the antenna is tuned to a specific frequency. In a scanner that can search 100 channels per second the tuner needs to be able to very quickly tune to different frequencies while still remaining accurate.

    Tuners often also have filters to keep out strong signals that may be in the area. For example, a scanner may have a high-pass filter in place that filters out signals below 30 MHz, because that's the lowest frequency the radio is designed to tune. Or, if a radio is only supposed to receive signals in the 144–148 MHz range they may put a band-pass filter in place that only allows frequencies between 144 and 148 MHz through. This will help keep out spurious signal caused by strong FM commercial radio stations or pager transmitter. FM radio stations and pager transmitters are both notorious for causing front-end overload in radio scanners.

    Some high-end radios will have multiple filters that are switched on and off depending on the frequency range the radio is tuned to. With scanners that need to receive signals over such a wide frequency range, it is extra challenging to make a front-end amplifier that works well across the entire range they are intended to receive.

3. *Mixer:* A mixer is sometimes referred to as a down converter. The signals from the RF amp and local oscillator enter the mixer. The mixer combines these waves and outputs a signal at a set frequency, referred to as the Intermediate Frequency (IF). No matter what frequency the radio is tuned to, the IF that comes out of the mixer is always the same. On most scanners the first IF output frequency is 10.7 MHz. Using an Intermediate Frequency makes it convenient to design the radio's components, because they can be built around specific IF frequencies.

In many radios this IF signal will actually go through multiple down converters and filters until it is at a frequency the demodulator can process. A radio with more conversions will filter out more "birdies" (see Sidebar). To keep the explanation simple, this diagram lumps these stages together, rather than break them out into multiple blocks.

WHY 10.7 MHZ?

The choice of 10.7 MHz for the Intermediate Frequency is a convention that was settled upon for a number of practical reasons. The standardization came about when Superhet receivers were first being made for the broadcast FM band in the US. 10 MHz was picked because any harmonics created by the mixer would fall outside the FM band (which is 20 MHz wide). The 0.7 was picked because the frequency spacing in the US was 2 MHz—since 0.7 is not a multiple of 0.2, any harmonic signal generated by the mixer would fall between channels, and thereby minimize their interference. From then on, 10.7 MHz became the popular value for other FM receivers such as scanners, most likely because there are lots of parts that are already designed and tuned to work with this IF, and using those parts is cheaper than designing new.

Most scanners put the signal through 2–3 conversions. This is referred to as double and triple conversion. Some high-end radios have quadruple conversion. Generally, the more conversion, the more likely the radio is to filter out spurious signals. These will work better in RF rich environments like urban areas. As radio emitters become increasingly commonplace, more and more places are becoming RF rich. Today most scanners are triple conversion, but this needs to be watched more when purchasing a used radio. We recommend getting a triple conversion scanner if you can afford it.

4. *Local Oscillator:* The local oscillator could be considered separate from the mixer, but it's a key component that makes the mixer work. The local oscillator creates radio waves at different frequencies, and is often referred to as a VFO (Variable Frequency Oscillator). The frequency created by the VFO changes in proportion to the frequency the radio is tuned to. This occurs so when radio waves from the local oscillator and tuner are combined they always exit the mixer at the IF frequency. If the VFO did not change frequencies, the IF produced by the mixer would change as the tuner frequency changes.

5. *Demodulator:* The demodulator extracts information carried by the radio wave, and (usually) converts it to an audio wave. For example, an FM radio uses an FM demodulator to extract the information needed to reconstruct the music the radio station is broadcasting. There are many types of modulation and demodulation. Other types will be covered later in this chapter.

6. *Audio amplifier:* This component amplifies the audio signal coming out of the demodulator to a level the end user can hear it. This is almost always variable, in order to provide volume control. The audio amplifier often contains filters to clean up the signal. Although the filters can make the

BIRDIES

"Birdies" are spurious signals and harmonics produced by the radio circuits. This is internal RF noise generated by the radio itself. Birdies make it appear like a signal is present where there really is not. A birdie will either be a signal that is silent, or will sound like static. To tell if you have tuned to a birdie on your scanner, remove the antenna. If the signal is still there, it is a birdie.

Most scanner manuals list the birdie frequencies for their radio. It is still good to verify this list, because new birdies could appear depending on slight difference in the manufacturing process. To find the birdies, take the antenna off your scanner and have it search all the frequencies it can find. Anytime it stops on a signal, it is probably a birdie.

If you are unsure, put the antenna back on. If attaching the antenna pulls in a strong signal, like a local pager tower, FM broadcaster, or TV station, then it is not a birdie—just a signal strong enough to receive without the antenna.

Since you're going through the trouble of identifying them, keep a list of these birdies for future reference.

When you setup your scanner to search for new signals (covered in the onsite profiling chapter), you can consult the birdie list to see if the signal you got was a birdie or not. Some people will also lock out the birdies so they are ignored during searches. Note that it's entirely possible for a real signal to be transmitted at a frequency that happens to be a birdie frequency of your scanner! Blocking out the birdies *could* cause you to miss a signal if it is on that frequency. Because of this, we do not recommend locking out the birdies until you've determined that the signals are indeed originating from inside your scanner.

Thanks to technological advances with equipment, finding a list of birdies is not required as much these days. If you have a triple conversion scanner, the number of birdies will be minimal. This is more important, but still not critical, on older double conversion radios.

TIP

The components that process the received signal from the RF amp to the first Intermediate Frequency are referred to as the *front end* of the radio.

audio sound better, they can cause problems if you try to decode data signals by feeding the audio output from a scanner into a computer. Oftentimes these filters will manipulate the signal enough that data signals, especially signals over 1200 Baud, cannot be recovered.

To get around this, it is necessary to pull the data signal before the audio is cleaned up by getting the audio from the discriminator output. The discriminator output provides access to the unfiltered audio signal. Some scanners have discriminator output ports built in, which makes it easy to access the unfiltered signal.

DIY RADIO MODIFICATION: DISCRIMINATOR OUTPUT

Scanners that do not have this convenient feature can be modified by opening up the radio and soldering an audio line at a specific point on the circuit board. This is a relatively easy modification, which can be performed by the casual hobbyist or enthusiast, in a few minutes for less than $5 in parts that can be obtained from Radio Shack or similar stores, using a soldering station, a Dremel tool for drilling holes, and a screwdriver.

Detailed instructions are beyond the scope of this book, as the procedure varies depending on the model. A Google search will often produce instructions on how to add a discriminator output to your scanner. Figure 2.8 shows the inside of a scanner with a discriminator output added. Of course, the usual disclaimers apply. This will void your warranty and we are not responsible if you break your radio!

FIGURE 2.8 Inside of a Scanner with a DIY Discriminator Output Added, Reprinted with Permission from Meagan Call

Filters

A filter is a device that allows certain frequencies to pass, and rejects other frequencies. Filters are used inside radios to clean up both RF and audio signals. Each filter has a cutoff frequency which is the frequency at which the filter reduces (attunes) the signal being passed through it and a decibel (dB) rating which tells how much the signals are reduced by the filter. A filter may not completely eliminate a signal if the input signal is stronger than the dB rating of the filter. Filters can also have a fixed cutoff frequency or a variable cutoff frequency. Classically, filters are electronic circuits. However, as more radios become software-based, more and more filters are being implemented in software.

There are two basic types of filters high-pass and low-pass filters. A high-pass filter allows frequencies above the cutoff frequency to pass through the filter (Figure 2.9). A low-pass filter allows frequencies below the cutoff frequency to pass through the filter (Figure 2.10).

Band-pass and notch filters are also common filters used in radios. A band-pass filter allows a set range of frequencies to pass and is created by combining a low-pass filter to remove signals below the target band and a high-pass filter to remove signals above the target band (Figure 2.11). Band-pass filters are helpful to eliminate strong signals outside of the target band.

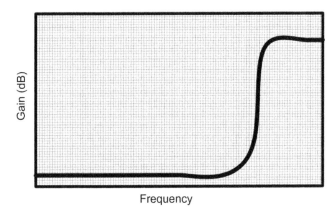

FIGURE 2.9 High-Pass Filter

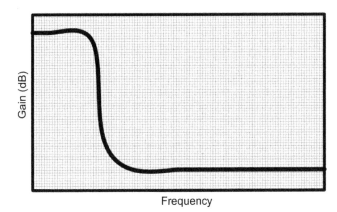

FIGURE 2.10 Low-Pass Filter

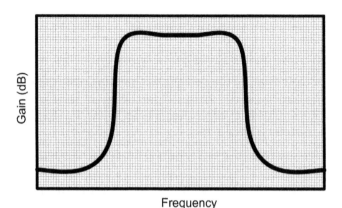

FIGURE 2.11 Band-Pass Filter

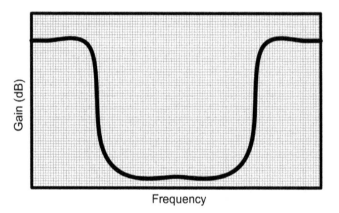

FIGURE 2.12 Notch Filter

Notch filter (Figure 2.12), also called a band stop, removes a section of the spectrum. Notch filters are helpful to remove a specific strong signal or band of strong signals that are causing interference. For example, commercial FM transmitters can overload the front end on some scanner. If this occurs placing a notch filter that attunes 88–108 MHz between the antenna and the scanner will fix the front-end overload.

ANTENNAS

Radio waves passing through the air are very weak. The antenna helps gather and strengthen the available signal in the air. When RF waves pass over an antenna, they induce their signal, which is comparatively weak, into the antenna that, which resonates with the frequencies it is catching from the air, boosting the signal, and feeds it into the radio.

The antenna is one of the most important components in any radio. A good antenna can easily mean the difference between detecting and missing a signal. In fact, given the choice, we would prefer a high-quality antenna paired with a low-quality radio than a low-quality antenna paired with a high-quality radio. The antenna is that important.

Antennas are used when signals are both transmitted and received. Because the focus of this book is on receiving signals, the next section will focus on how antennas receive signals and ways to improve how antennas can receive signals. Similar techniques can be applied to transmitter antennas to improve their performance.

Antenna Theory

A number of concepts are key to understanding antennas:

1. *Resonance:* Resonance occurs when an electric signal travels from one end of a wire to the other and back, in the same amount of time as the period of one cycle of the RF frequency. In other words, when the length of the wire is equal to half the wavelength. RF signals will resonate with antennas with lengths at multiples of the wavelength, as well, a phenomenon known as *harmonics*. Harmonics makes it easier for the antenna to pick up signals of the resonant frequency. Full wavelength antenna can get very long which makes them unwieldy when mounted on a handheld radio or vehicle. Because of this people often use 1/2 and 5/8 wave antennas. To understand why 1/2 and 5/8 wave antennas function, it is essential to understand resonance. The more resonant an antenna is with respect to a given frequency, the less RF energy is required to excite the antenna and allow the antenna to pass a signal at that frequency through to the radio. The *low threshold*, or how weak a signal the radio is able to pick up, determines the sensitivity of the radio. This is typically measured in microvolts (μV). RF energy passing over an antenna excites the antenna, generating a very small voltage, also measured in μV. Antennas can also act as *filters*. Tuning an antenna to resonate at specific frequencies will make it easier to hear signals on that frequency. Note, however, that tuning an antenna to resonate on specific frequencies will make it more difficult to hear other

---

TIP

Bigger isn't always better, especially when it comes to antennas. In fact, the shorter the antenna, the better it will hear higher frequency signals, since high frequency means short wavelength.

frequencies. This is fairly easy to do with a telescoping whip antenna by changing the length of the antenna. Remember longer isn't always better. Sometimes you will need to shorten an antenna to hear higher frequency signal.

It is often useful during onsite reconnaissance to be able to determine the approximate frequency of a transmitter by the length of its antenna. Knowing this length will help you zero in on the frequency ranges that interesting signals are more likely to be found, and decrease the time it takes to determine the exact frequency used by the target. The required length of dipole and whip antennas is determined by the following equation:

$$\frac{492}{\text{Frequency (MHz)}} = 0.5 \times \text{wavelength} = \text{length of antenna (in feet)}$$

To obtain the length in inches, multiply by 12. For the length in meter, use this equation instead:

$$\frac{300}{\text{Frequency (MHz)}} = 0.5 \times \text{wavelength} = \text{length of antenna (in meters)}$$

Multiply this result by 100 (or simply divide 3/f) to obtain the length in centimeters.

- *Polarity:* An antenna's polarity determines which spatial axis the antenna is most responsive in. There are two types of polarity: horizontal and vertical. Understanding the polarity of the antenna you are using will help you to position it properly, to get the maximum gain, and to aim it (if the design of the antenna necessitates) at the transmission source. Antennas with horizontal polarity are most sensitive in a horizontal plane, and properly deployed the antenna should look like it is laying down, like the old style television directional antennas (Figure 2.13) you may still see on rooftops. Antennas with vertical polarity should stand vertically, as with a whip antenna on a portable FM radio or CB radio. These are also the most common antenna for RF scanners.

  To get the best performance from your antenna, you want to match the polarity of the transmitting and receiving antenna. So for most systems in this book the antenna on the receiver should be held vertically to get the best performance.
- *Antenna gain:* The amount that an antenna increases the signal strength is called gain, which is measured in decibels (dB). The decibel scale is logarithmic rather than linear, and due to this, just a few dB of gain can make a significant difference. As an example, the difference in signal strength between a 50 W light bulb and a 100 W light bulb is 3 dB.

FIGURE 2.13 Horizontal Polarity TV Antenna

Signal Strength

There are three basic ways to increase signal strength when receiving a signal: amplification, antenna tuning, and antenna orientation. Using an amplifier will amplify noise as well, including those you may not want, and may simply result in a louder version of the same noisy signal. You can also tune the length of the antenna to the frequency of interest. This can be done by adjusting the length of the antenna, or changing to an antenna tuned for the range you are interested in. Finally, using a directional antenna that is designed to focus the signal energy can increase the signal strength.

Antenna Diagrams

To understand the reception properties of a given antenna, refer to the antenna diagram for the type of antenna. An antenna diagram is a graph, showing the gain and radiation pattern an antenna has around it. This is useful for determining the characteristics of a directional antenna, or to see how uniform an omnidirectional antenna is. No antenna is going to perfectly receive in all directions, so a diagram can help you understand the characteristics of the antenna. Antenna diagrams are useful for understanding the concept of directional antennas. By studying antenna diagrams, you can see the lobes generated by making an antenna directional. This is an important concept to understand if you plan to use antennas for direction finding.

> ### TIP
>
> When trying to determine the direction of a signal source, change the orientation of the antenna until you find the orientation which allows you to receive the signal most clearly and strong. As you do this, you may see a small jump at a sub-lobe in the antenna's gain field, which has the potential to give the appearance that the direction has been found, leading you on a wild goose chase. Knowing the antenna diagram for the antenna you're using will help you to avoid being fooled. Always look for the most gain when orienting the antenna toward a signal source.

Popular Types of Antennas

Broadly, antennas can be divided into two groups: directional and omni-directional. There are many different types of antennas—dozens or hundreds, depending on how you classify them. This section will focus on just the types that are of interest to penetration testers performing wireless reconnaissance.

Omnidirectional and Directional Antennas

Omnidirectional antennas receive signals equally from all directions. Directional antennas pull in signals better from one direction. In this direction, they can detect a weaker or more distant signal than an equivalent omnidirectional antenna. The trade-off is that they do this by decreasing their ability to pull in signals from other directions.

> ### NOTE
>
> **How Omni is Omni?**
> While the prefix omni—implies that the antenna is able to receive signals from any direction, technically omnidirectional antennas are usually only omnidirectional in a single plane. For example, depending on its orientation, an omnidirectional antenna may detect signals to the North, South, East, and West, but not above or below.

> ### NOTE
>
> Gain: The difference in signal strength achieved by an antenna is known as *gain*, which is measured in decibels (dB). Directional antennas adjust the gain pattern to better receive signals from a specific direction. A well-designed omnidirectional antenna can also provide gain on specific frequencies the antenna is tuned to.

Directional antennas have both positive and negative gain, depending on from which direction you are looking at the antenna. This is mapped out in antenna diagrams.

Types of Omnidirectional Antennas

- *Discone:* Discone antennas are broadband antennas (able to receive a wide range of frequencies) that are generally base mounted. There are a few commercial discone antennas available for handheld radios, the most popular of which are tuned for cellular phone frequencies. Discone antennas can be designed and built to be sensitive across a wider range of frequencies than other types of antennas.
- *Whip:* A very common type of omnidirectional antenna is the whip antenna (Figure 2.14). Typical scanners come with a whip antenna. These antennas are usually so inefficient that they have no gain, however this

FIGURE 2.14 Whip Antenna, Reprinted with Permission from Meagan Call

is a sacrifice made to allow the antenna to cover a wide frequency range. This wide range also means that it is also possible that the antenna will have a negative gain on certain frequencies. Usually, they are made to operate best on a specific frequency largely based on the length of the antenna. Telescoping whip antennas can be tuned to a variety of frequencies based on how much the antenna is extended.

- *Dipole:* Antenna made up of two wires connected in a straight line. One wire connects to the radio and other wire connects to group. Usually made to operate best on a specific frequency. Commonly used as a base antenna. The biggest advantage to dipole antennas is they are a very simple design and so are cheap to make. Additionally because they are a simple design they are very easy to make at home and are usually the first home-brewed antenna people make.

Types of Directional Antennas

- *Yagi:* This is the familiar roof antenna used for television reception (Figure 2.15), which were a common sight in the golden age of NTSC broadcast, before cable and satellite television caught on and eventually largely replaced it. It consists of a central spine with numerous "ribs" of various lengths sticking out laterally.
- *Panel and wave guide:* These types of directional antennas (Figure 2.16) which may be familiar to penetration testers from attacking wireless

FIGURE 2.15 Yagi Antenna

FIGURE 2.16 Wave Guide Antenna, Reprinted with Permission from Meagan Call

802.11 networks. The "cantenna, " an inexpensive hardware hack involving a Pringles potato chip can, was a well-known example of this type of antenna, which happened to have dimensions and materials suitable for picking up Wi-Fi signals. They are not of much use in the frequency rangers usually covered by radio recon (below 1 GHz).

MODULATION

Modulation is the method used to encode voice or data that is transmitted by a radio. Broadly speaking, modulation may be classified as analog or digital. Analog is the older of the two, but digital is already well established, and continues to expand its share of the market, due to its numerous advantages. Digital formats offer higher audio quality and allow data to be sent more reliably, can use data compression, and sharing of the media, which allows for greater transmitter density, and more users of a given frequency in a given area.

To properly listen to a signal, it is necessary for the transmitter and receiver to use the same modulation type. On some scanners, the modulation type is called the mode.

Analog Modulation

While most people are familiar with Amplitude Modulation and Frequency Modulation (AM and FM), there are many other modes in which different types of radios operate. FM is the most common mode of modulation in the United States that is of interest to penetration testing, but it is good to know and understand the other modes.

Common Analog Modulation Types

- *AM—Amplitude Modulated:* The strength (amplitude) of the radio wave is varied to encode the information into the carrier wave. Used with frequencies below 30 MHz, for shortwave radio transmissions, in the commercial AM band, and for aircraft communications such aircraft-to-aircraft or aircraft-to-tower.
- *FM—Frequency Modulated:* The frequency of the carrier wave is changed to encode information into the radio wave.
  - FM Narrow (FMN) is commonly used by two-way radio systems. It is designed to transmit low audio fidelity signals and takes up less bandwidth then the FM Wide (FMW) modulation (see below), which is used by broadcast FM stations and is discussed next. FMN is the most popular format for the focus of this book. With the exception of AM being used by aircraft, FMN is used for all two-way analog voice traffic transmitted on frequencies over 30 MHz.
  - FM Wide (FMW) is used by commercial FM Broadcasters because it can transmit a higher fidelity signal than FMN. This increase in audio fidelity comes at a price and FMW takes up more bandwidth then FMN.

- *SSB—Single Sideband:* This type of modulation is generally only used with shortwave frequencies below 30 MHz, because it makes better use of bandwidth. It is used for two-way communications, typically applications like ham radio and ship-to-ship communications. Broadcast shortwave and AM radio stations do not use SSB because it requires a more complex radio which can be more difficult for the end user to tune.

TIP

If you are planning on obtaining a radio that covers shortwave bands, consider getting one that supports single sideband so you can listen to more than just commercial shortwave.

Digital Modulation

Digital modulation types are becoming more and more common as people look for new ways to conserve bandwidth. Thanks to data compression algorithms, a digital signal carrying voice data can take up less RF bandwidth then an analog voice signal. Digital can be higher quality for voice traffic, as well. This is due to error checking, which keeps the 1's and 0's of the digital stream preserved and automatically corrected when the signal is distorted within limits. A distorted analog signal, by contrast, cannot be corrected this way at the receiving end, and therefore will sound distorted. However, this comes with a trade-off. A badly distorted analog signal may still be demodulated and be at least discernible as voice, and a good ear can make out intelligible speech through the static and distortion. With a digitally modulated signal, if the distortion is too great, the error correcting cannot restore the signal, and if a packet cannot be demodulated, it gets dropped, resulting in cutouts of the voice stream. Once you get enough dropped packets, you lose all voice traffic.

Currently, the great challenge with digital modulation is finding a radio or software that is able to decode it. The situation that exists today is similar to the "codec hell" issue that drove computer users crazy a number of years ago when trying to watch videos downloaded from the Internet. Each video file was encoded with its one of dozens of custom codecs. In order to decode the video data, you needed the right codec to watch the video, which was not included with the video file itself, and had to be downloaded and installed separately. Without any automated means of finding and installing the codec files, it could sometimes take a great deal of searching to find a web site that offered the codec for free download. Until just a few years ago, many video codecs in wide use were often license—and/or patent-encumbered, and had to be purchased in order to be used legally.

With digital radio demodulation, it's more a matter of proprietary modulation not being made available for scanners. Many manufacturers have developed their own method of modulation and consider it a trade secret that confers them an advantage if they keep it to themselves, forcing customers to buy from them. Oftentimes, the only way to demodulate a signal, other than with the proprietary radio that supports the format, is a time-consuming reverse-engineering effort.

TIP

During the offsite profiling stage of your penetration test be sure to check the http://www.
radioreference.com database for voice codecs commonly used by your target to verify if your
scanner can monitor them or research ways to decode the traffic.

APCO P-25

There are an ever-growing number of digital modulation types used by wireless systems. However, currently the only standard that can be decoded by consumer grade scanners is the P-25 standard. Other digital modulation types are covered in Chapter 9.

Project 25 (P-25) is an open standard made by the Association of Public-Safety Communications Officials (APCO) to promote interoperability between public safety radio systems. Many advanced radio features such as digital modulation and trunking are proprietary features created by manufacturers. These features traditionally did not interoperate with other manufacturers' equipment. This often made it difficult to coordinate the response to a large-scale disaster requiring multiple agencies or municipalities to respond because the radio system used by one city or agency would not be compatible with radios used by another city. The APCO P-25 standard was created to overcome these interoperability issues. The technical specifications for P-25 can be found in the ANSI/TIA-102 series of documents.

P-25 transmissions can also be encrypted using a variety of standard encryption algorithms such as Data Encryption Standard (DES), Triple-DES, or Advanced Encryption Standard (AES). A number of NSA generated encryption ciphers are also supported such as ACCORDIAN, BATON, Firefly, MAYFLY, and SAVILLE. However, few organizations implement encryption because of the additional hardware costs, administrative overhead of maintaining and distributing encryption keys, and poor reputation for performance encrypted radio systems have.

The P-25 standard is being deployed in a number of phases. Each phase adds additional features. Phase 1 is currently in wide deployment. At the time of this

NOTE

Older voice encryption systems had a reputation of degrading the sound quality and reliability of a radio system. Although these challenges have been overcome by modern day voice encryption systems, many users still avoid encryption because of this bad reputation.

writing, most P-25 systems were Phase 1 systems. Currently, multiple scanners can decide P-25 Phase 1 traffic. In Chapter 7 we discuss how to select a scanner suitable for monitoring these systems.

At the time of this writing, a half dozen P-25 systems had migrated to Phase 2 or Motorola's X2-TDMA system. Motorola's X2-TDMA system was released before the Phase 2 requirements were finalized and is largely based on the Phase 2 standard. Currently scanners support for Phase 2 systems is very limited. The GRE PRS-800 has experimental support for P-25 Phase 2 systems. As these systems become more widespread, most likely additional scanners will be made that support this system.

Common Types of Spread Spectrum Modulation

These modulation types all incorporate methods that spread a signal over a chunk of the spectrum to make better use of the bandwidth and avoid interference. Monitoring these systems can also be difficult, since the hopping pattern can be difficult to follow. Spread spectrum modulation enables sharing of the electromagnetic media, in other words more users in the same area can transmit in the same frequency range without interfering with each other. This is because the digital modulation broadcasts data in packets, and exercises a protocol to share tiny slices of time on the same frequency band with other radios that may be operating in the same area. In some systems, the hopping pattern is specifically made to be hard to track as an added security mechanism. Direct-Sequence Spread Spectrum (DSSS) and Frequency Hopping Spread Spectrum (FHSS) are two common types of spread spectrum modulation.

DSSS spreads the signal over the full bandwidth of the transmitter's frequency range. This is done using a pseudorandom sequence to determine how data is spread throughout the frequency range. DSSS signals look very much like white noise when looking at their RF graph. DSSS modulation is used by a variety of wireless systems such as GPS, CDMA-based cellular networks such as Verizon and 802.11b.

FHSS rapidly switches the radio carrier among many different channels in a pseudorandom sequence. FHSS is used by many consumer devices that operate in the 2.4 GHz band such as baby monitors, cordless phones, and wireless video cameras.

With both DSSS and FHSS the transmitter and receiver need to share the same pseudorandom pattern if they are going to exchange data. This offers a level of security because an attacker would need to figure out the pseudorandom pattern to intercept the transmission.

RADIO SYSTEMS

A single radio by itself is not of much use. In order to work, you need a system. The most basic system consists of a transmitter and a receiver, using a common modulation type. More commonly in radio reconnaissance, we deal with two-way radios, from pairs of transceivers to complicated radio networks and trunked systems. Next, we'll review the common types of systems and describe how they are structured.

Simplex and Duplex

Two-way communication can be handled in a few different ways, depending on the sophistication of the radio system. This gives different systems different capabilities.

Simplex

Sometimes simplex is referred to as half-duplex. In simplex two-way radios, communications take place over a single frequency. This means that only one person can talk at a time. To prevent talking over each other, most users of simplex systems employ a spoken protocol to mark beginning and end of transmission, (e.g. saying "Over" when done speaking.) Monitoring the conversation on a simplex system is easier, because you only need to monitor one frequency.

Two-Frequency Simplex

Some call it split-frequency simplex, or split-frequency half-duplex. Transmission and reception take place over different frequencies, but you can only transmit OR receive at the same time—you cannot do both. If listening in, in order to hear the entire conversation you will need to listen to both frequencies on your scanner.

Duplex

Sometimes called full duplex, to differentiate from half-duplex. With a duplex radio, two frequencies used—transmit and receive. The radio is able to transmit and receive simultaneously, allowing both parties to speak and listen at the same time like a telephone conversation. In fact, a common application for duplex radios is with cordless telephones. With duplex radios, usually it is necessary to monitor both frequencies to get the entire conversation. However, some systems retransmit both sides of the conversation over one of the frequencies. Transmitting both sides on one channel is common with cordless phones. So always beneficial to try to monitor both frequencies to see if one retransmits both sides. If one does, it saves you the work of having to monitor the other.

If you are only hearing part of a conversation on your scanner you are probably listening to a two-frequency simplex or full duplex conversation, so this should be a clue that you have another frequency that you need to find. Once you've found both the send and receive channels, program both frequencies into your radio to monitor them.

Repeaters

A radio repeater is exactly what the name sounds like. It is a device used to extend the range of a radio. When a signal is transmitted into a repeater, it boosts the signal and rebroadcasts it, extending the range of the original signal. It is essential to note that repeaters have separate input and output frequencies. Were it not for this, the retransmission would be picked up again by the repeater, creating a feedback loop, and render the repeater useless. When listening to repeater traffic, be sure to listen for the output frequency; if you listen to the input frequency, you will only hear the traffic in your local area. For amateur radio repeaters there are set repeaters offsets used depending on the band of the frequency the repeater operates on. Commercial repeaters do not follow a standard. Instead the input and output frequency are depending on what frequency the FCC licenses for them to use.

Repeaters may be fixed, or they may be mobile. Fixed repeaters are (usually permanently) mounted in a fixed location, while mobile repeaters are attached to a mobile platform, such as a vehicle.

Fixed repeaters are also often stationed in high places to increase the broadcast area by extending the horizon and minimizing line-of-sight issues. They are commonly mounted on hill or mountain tops, or the highest available areas of local geographical elevation, tall buildings, and purpose-built towers. For example, if a handheld radio is being used in a valley, a repeater positioned on high ground can take the signal and retransmit it so that other can receive it. Multiple repeaters can be linked together so a signal is transmitted over multiple locations, attaining even greater coverage. In this way, short-range signals can be relayed across vast distances, even across a country or state. Mobile and fixed repeaters may be used in concert, augmenting each other, to create radio networks.

Repeaters frequently use Digital-Coded Squelch (DCS) and Continuous Tone-Coded Squelch System (CTCSS) (see Media Access Control in Radio, below) to prevent spurious signals from entering the repeater and being retransmitted. This can also function as a simple and inexpensive form of access control because someone who wants to transmit using the repeater needs to know the CTCSS or DCS code used by the repeater. However, keep in mind that many radios have a feature that will automatically determine the CTCSS or DCS code in a received signal so this is also a fairly weak form of access control.

> **NOTE**
>
> Some police vehicles act as repeaters for handheld radios, repeating and amplifying the signal from the officer's handheld, allowing the signal to reach the main dispatch center.

Media Access Control in Radio

Media access control is any of the types of methods that enable multiple people to share the RF spectrum. A number of systems have been developed over the years to make better use of the spectrum.

The radio spectrum is a finite commodity, and getting increasingly crowded. Because of these limited availability of open spectrum, it is expensive to get licenses for multiple channels. To deal with these limitations, different approaches, from techniques improvised in the field, to increasingly sophisticated technological innovations have been developed to provide ways for multiple groups to share the same frequency of system.

In its simplest form, operators simply listen to the channel and wait for it to be clear before transmitting provides access control. More complex access control consists of using CTCSS or Digitals Squelch Tones (see below). In its most complex form, access control is accomplished with trunking.

CTCSS

Continuous Tone-Coded Squelch System (CTCSS), also known as tone squelch, is designed to allow users on a shared frequency to hear only users in their user group. Each group is assigned its own squelch tone, and the radio only plays to the audio when the squelch tone is transmitted. The transmitter adds a unique sub-audible code to the transmission. The receiver, if in CTCSS mode, listens for the transmission and the unique sub-audible code. The receiver's audio will only activate if both the transmission and the sub-audible code are present. Theoretically, CTCSS will allow a user to hear only transmissions by those in their user group, and not be subjected to the transmissions of others on the same channel.

CTCSS-equipped receivers generally can operate in either CTCSS mode or normal mode. While in CTCSS mode, the receiver's audio will only be activated if the transmitted signal was sent using the same CTCSS tone. CTCSS tones are standardized by the Electronic Industries Alliance (EIA) in standard RS-220.

One important caveat with CTCSS systems, from a user perspective, is that CTSS does not somehow magically create extra bandwidth. There is still only one channel in use, and it cannot carry more than one transmission without

> **NOTE**
>
> CTCSS does not provide any scrambling or privacy protection. Heeding the CTSS signal and ignoring the transmission from other groups is "voluntary" and a radio does not need to know the code tone in order to listen to the broadcast.

> **NOTE**
>
> CTCSS has several different commercial names that vary by manufacturer.
>
> - Motorola—Private Line or PL Tone
> - GE and Bendix—Channel Guard (CG)
> - Icom—C.Tone
> - Kenwood—Quiet Talk (QT)
> - Johnson—ToneGuard (TG), CallGuard (CG)

interference. However, because users in one group cannot hear the transmissions of the other group or groups, they may not know whether or not the channel is clear before attempting to transmit. Because of this, most radios using CTCSS are designed so they automatically will not transmit when the channel is occupied. However, the radios in these systems are not capable of prioritizing traffic based off of user groups. So, if a high-priority user like a fire fighter needs to transmit but a low-priority user like a dog catcher is talking, the fire fighter will need to wait for the dog catcher to finish before they can transmit. Obviously, for this reason, CTCSS may not be suitable for mission or life critical applications. As more users, or in this case, more groups, use the same channel, congestion builds as traffic increases. The more often someone is transmitting, the more likely there will be interference. There are some ways to mitigate this interference, including features that do not allow new transmissions while the channel is in use.

Reconnaissance of CTCSS Equipped Radios

A receiver with the CTCSS function turned off will hear all transmissions on the channel. Users of CTCSS may incorrectly believe that CTCSS provides some degree of security, and may be more likely to discuss non-public information. This can of course be used to our advantage as penetration testers.

Digital Code System

Digital Code System (DCS) is a digital version of CTCSS. Motorola called their version of this access control technology Digital Private Line (DPL).

> **NOTE**
>
> CTCSS and DCS tones are also used by repeaters to prevent rebroadcasting spurious signals. In these instances the repeater will only rebroadcast signals that have a specific CTCSS or DCS tone transmitted with the input signal.

For DCS there are only up to 100 possible codes. As with CTCSS the presence of a DCS tone does not encrypt or scramble the signal, so it's not really very private. As with DCS, a receiver with the DCS function turned off will hear all transmissions on the channel. Again, the false sense of security can provide opportunities for attackers and penetration testers should look for them.

Trunking

During the early years when radio was still a new technology, and not widely deployed, efficient use of the spectrum was not a major concern. Those days are long gone, and the RF spectrum is a limited commodity. In the early days of radio, the RF spectrum was a frontier being settled, and the FCC was brought into existence to apportion the spectrum to avoid conflicts. Later, as radio became increasingly popular, even this did not adequately apportion the band—it wasn't enough to license one operator to use a particular frequency in a particular region at a particular power level, especially with bands of the spectrum that are not used for commercial broadcast, and thus are not in use 24/7. For two-way radio systems with many users in the same general region, licensing in the old way was no longer adequate. It became necessary to devise better techniques of utilizing those parts of the band, in order to allow for more efficient use of it by many operators. And this was the need that trunking was devised to address. Where older systems allowed only a smaller number of operators to use two-way radios in a given region without constant chatter making the system difficult to use, trunking systems made it possible to share that band more efficiently, allowing more users in a narrower band of channels. Today, the radio spectrum is a finite resource, and there are increased demands on and competition for access to the RF spectrum. The RF bands used by two-way radio systems is in demand too much for underutilization to be acceptable. With the FCC tightly controlling access to frequencies, and step sizes growing smaller to fit as much traffic as possible into frequency ranges, efficiency is a must. This is where trunked radio systems come in. A trunked system is designed to make better use of the spectrum to allow a greater number of users to share a small number of channels, in greater density than otherwise might be practical to achieve.

Trunked Radio Systems in Depth

Trunked radio systems centrally manage a pool of channels, and intelligently switch users to whatever channel is open at a given time. This is often a difficult concept for scanner enthusiasts to understand, as trunking represents a paradigm shift in design of the system. Trunked radio systems are one of the most complex types of radio systems in use today. Trunked radios use several channels or frequencies, and allows those channels to be shared by a large number of users, in multiple talkgroups, without their conversations interfering with each other.

Talkgroups are defined groups of users. For example, all police officers and dispatchers would belong to a police talkgroup; all fire fighters and fire employees would belong to a fire talkgroup, and all municipal waste collectors would be in a waste management talkgroup.

Trunked systems use a control channel, called the trunk, which transmits data packets which allow a talkgroup to carry on a conversation by telling members of a talkgroup which frequency to communicate on when they key up. This allows for a large number of users to communicate using only a small number of frequencies, and more efficient use of those frequencies.

Trunked radio systems operate on the assumption that not all talkgroups will be in use at once. This allows for the channels to be shared by multiple talkgroups, achieving a more efficient use of the band. Rather than requiring all radios to be set to same frequency, thus monopolizing that frequency whether or not it's actively in use, trunked systems allow for many users to share multiple frequencies. This means that far fewer frequencies are required to support the radio traffic, and therefore greater efficiency of spectrum allocation.

Talkgroups also allow for more granular assignment of user groups. This keeps conversations relevant to the group that needs to hear them. Consider a simple example of municipal refuse collection. In a traditional non-trunked system, the refuse collection department would be assigned a frequency, and all users would hear all transmissions. With talkgroups, several distinct groups can be created. There could be one talkgroup for all trash collectors, and another talkgroup for an overlapping subset of trash collectors such as those who only pick up recycling. When a communication is only relevant for the recycling group, the recycling talkgroup can be used, sparing the wider refuse collection group from having to hear the communication.

Trunked systems generally utilize a bank of channels for talking, and a control channel of some type. Depending on the type of system, the control channel may be either fixed (dedicated), or a random channel which changes (dynamic). Regardless of the type of trunking system and the control channel type, all trunked systems need a way to let users know on which frequency their talkgroup is on. The control computer sends a signal to all the radios which are part of the talkgroup, which instructs the radios assigned to that talkgroup to tune to a specific frequency.

In many, if not most, trunked systems, the transmission and reception will remain active on the same frequency until the communication sequence is complete. With these types of systems, even without a trunk-capable radio at your disposal, you can at least follow the conversation if you should come across one. Keep in mind, however, that without trunking support, it won't be possible to know which talkgroup you have found, and you are in essence in the dark.

Other systems will change the frequency used for communication *each time* a user keys up. In these types of systems, it is generally very difficult, if not impossible, to follow a conversation unless you have a trunk-capable radio.

An additional advantage of trunking systems is that, should the allotted frequency become saturated, the controller can prioritize traffic. This is especially useful in municipal systems where there is limited spectrum available for multiple services. For example, in a given city, the Fire Department may have Priority over the Police Department, who may have priority over Animal Control, who may have priority over the Service Department, and so on.

Many trunked systems also have a feature where the radios (mobile, handhelds, and central dispatch) have an emergency button. When the emergency button is pressed, the system will drop lower priority users and traffic from the system to allow the emergency traffic to get through.

Manufacturers of Trunked Radio Systems

There are five major types of trunked communication systems in wide use in the United States: Logic Trunked Radio (LTR), developed by the E.F. Johnson Company; Enhanced Digital Access Control System (EDACS), developed by General Electric (GE); Motolora's Type I and Type II systems and P-25 which is an open standard developed by Association of Public-Safety Communications Officials (APCO). With the exception of P-25 all of these trunking systems are made by competing companies so the trunked systems are proprietary and not compatible with each other. The most popular system is probably Motorola's. LTR is used primarily in single-site applications. P-25 systems are quickly gaining in popularity because it is an open standard that offers interoperability and the ability to use multiple manufacturers' radios in the same system.

The larger and more complex the target is, the more likely it is that they would have deployed advanced radio systems. Trunked systems are common in government, as they allow municipal services such as fire, police, and service departments to share an allotted spectrum. In the commercial sector, trunked systems are deployed at large corporate campuses, amusement parks, and sporting arenas. Increasingly, companies are also lease space on an existing trunked system, instead of building out their own systems.

Monitoring Trunked Radio Systems

Listening in on trunked systems is more complex than listening to non-trunked systems. The penetration tester is likely familiar with packet switching in computer networks, and the operation of the trunked radio system is essentially like that of a packet switching network. There are scanners specifically designed to listen to trunked conversations, and you will need one to effectively listen to trunked traffic. Without a trunk-capable scanner, you may be able to pick up bits of conversations; however it is nearly impossible to consistently get the entire conversation. Selecting and the basics of using a trunked-capable scanner are covered in Chapter 7.

Trunking systems that need to cover a wide area or multiple sites can be made up of multiple repeaters. Many states have created trunked networks using multiple repeaters that cover the entire state. In trunking terms each repeater or transmitter tower is referred to as a site. Multi-site trunked systems can be divided into two categories simulcast and SmartZone.

In a simulcast system a transmission that comes into the systems is broadcast by every site in the system. In a SmartZone system each unit checks into the nearest site in range so the sites have a list of all the units using them at any given moment. A SmartZone site only broadcast traffic if a user in the target talkgroup is using that specific site.

An analogy that is helpful to computer users familiar with Ethernet networks is to image a multi-site trunked radio system as an Ethernet hub or switch were each port on the hub or switch is a trunking site. In this analogy a simulcast system would be an Ethernet hub because traffic that comes into one port (site) is broadcast out to all the other ports (sites) in the system, it does not matter which talkgroup the traffic was destined for. All ports (sites) still see the traffic. A SmartZone system can be thought of as an Ethernet switch where traffic that comes in from one port (site) destined for a specific talkgroup is only broadcast to the ports (sites) where members of that talkgroup are connected.

TIP

Trunking capable scanner often has complicated programming functions. The best reference for programming your scanner to work with trunked systems is your scanner owner's manual.

As you can image, simulcast system make very poor use of valuable channels in large and complex systems. If you have a trunked system that spans over a state oftentimes there is no reason to broadcast police or fire traffic from one city to all the cities in a state. SmartZone systems make much better use of the limited number of channels in a large trunked system but only transmitting information if a user in that site is in the target talkgroup.

From a wireless reconnaissance point of view, the types of system used will affect what traffic you can hear, based off of which sites you can hear traffic from and which end users are using those sites. In a simulcast system you will hear all traffic in the trunked system no matter which tower you can monitor. In a SmartZone system you will miss traffic if there are no users of the target talkgroup checked into a tower that is within range. The easiest way to avoid this is to be within range of the site that is closest to your target. Most likely the talkgroup you are interested in will be in use near your target so units in that talkgroup will be checked into that site.

SUMMARY

These fundamental concepts that we have explored in this chapter are essential knowledge for the penetration tester. Refer back to this chapter often to develop familiarity with these concepts. To review what we've covered:

- Radio Frequency (RF) is a large part of the electromagnetic spectrum, from 3000 kHz to 300,000 MHz.
- All electromagnetic signals move through a vacuum at the speed of light. Because of this, frequency and wavelength are inversely related to each other. Low-frequency waves have a high wavelength, and high-frequency waves have a short wavelength. If you know the frequency, you can determine the wavelength and vice versa.
- The SuperHeterodyne design is the most common in modern radios. It consists of an antenna which feeds incoming signals into a "head end" (comprised of the RF Amplifier, Tuner, and Mixer), and from there is fed into the Demodulator for conversion, amplified by the audio amplifier, and finally output through a speaker.
- A good antenna is critical to radio performance. The antenna's length determines what frequency it is tuned to; shorter antennas will pick up higher frequency waves. The wave length is twice the length of the antenna. To know what length an antenna is tuned for what frequency, use the equation

$$\frac{492}{\text{Frequency (MHz)}} = 0.5 \times \text{wavelength} = \text{length of antenna (in feet)}$$

- FM is the most popular analog modulation type above 30 MHz, and is of the most interest to penetration tests for wireless recon.
- Digital modulation types are becoming more common. This is a challenge, given that scanners currently have very poor support for digital formats.
- Currently P-25 is the only digital modulation supported by scanners.
- Simple radio systems consist of a transmitter and one or more receivers, or more commonly with the types of radios that are of interest to penetration testers, multiple transceivers.
- Repeaters allow signals to be transmitted over a wider area by amplifying and rebroadcasting the input signal. Be sure to monitor the output frequency of a repeater in order to receive all the traffic.
- Trunked radio systems are advanced networks that allow for multiple talkgroups to share a small band of channels, controlled by a trunking master, which acts like a switchboard to route transmissions to the appropriate listening party and conserve radio bandwidth.

FURTHER LEARNING

Beyond what can be covered here, there is a great deal that cannot be adequately covered in a single chapter. Once you have a grasp of the fundamentals covered here, be sure to continue your education. We recommend a few next steps:

1. *Get your amateur radio license:* American Radio Relay League's (ARRL) web site (http://www.arrl.org/) is a great source of information on how to get your amateur radio license.
2. *Further reading:* The American Relay Radio League (ARRL) puts out a lot of good books. Two that stand out are: *Basic Radio: Understanding the Key Building Blocks* by Joel R. Hallas and *The ARRL Handbook for Radio Communication.* The Handbook is a phonebook-sized tome, and is updated annually. Older editions can be found inexpensively secondhand. They do add and revise content each year, but much of the information in the old editions is still useful. Another very good read is *The Art of Electronics* by Paul Horowitz and Winfield Hill. This is not specific to radio, but will provide good background for general electronics which is important to radio.
3. *Online courses:* Many schools, notably the Massachusetts Institute of Technology, are releasing courses online for free. This is a great way to learn about these topics from top notch institutions for free. The MIT open courseware web site is http://mitx.mit.edu/. A good course to look into is 6.0002.x: Circuits & Electronics.

   iTunes U [http://www.apple.com/education/itunes-u/] also has a number of recorded lectures if you search for radio and electronics courses.

Targets

As with any other type of reconnaissance or penetration test, it is vital to the success of radio reconnaissance to define the targets prior to beginning. The end goal of radio reconnaissance, for the purposes of this book, is to gather information that will enable either physical or logical penetration.

There are two basic types of targets. The first is two-way radios used for verbal communication, and the second type of targets is devices that use radio frequencies to transmit audio or video information.

CONTENTS

TWO-WAY RADIOS USED FOR VERBAL COMMUNICATION

Radios used for verbal communication have obvious applications for the penetration tester. The best example of these types of radios is the walkie-talkies. Monitoring the guard radios or other radios used by the target organization's staff will often allow the penetration tester to know where guards and staff are physically located, and help the penetration tester avoid detection. Monitoring these communications can also give you insight into company lingo, employee names, and company culture. Be aware that it isn't just guards who use radios to communicate. Often times other staff members will use radios to communicate with each other. When out and about, look around for radios. You may be surprised to see employees of retail establishments, restaurants, and other places frequently using radios to communicate. This is in addition to maintenance staff as well as others. Remember, it isn't just guards who can provide you with valuable reconnaissance information or discover you and call the police during a physical penetration test.

DEVICES THAT USE RADIO FREQUENCIES

Radio waves are of course invisible to the human eye. That is a good thing, otherwise our vision would be completely obscured by the multitude of radio waves passing by and through us at all times.

Radio is everywhere and is commonly used by electronic devices. As a society, we have been "cutting the cord" in many aspects of our daily life and work. Cameras, headsets, telephones, and many other devices used to require a cable to transmit their information. Now, of course, many, if not most, of these devices transmit wirelessly using radio frequencies.

Ways to use the information gathered from these targets will be covered in later chapters.

Walking into any organization, one of the first people you encounter will often be a receptionist, and receptionists often use wireless headsets. Of course, many other employees aside from receptionists use wireless headsets. These headsets are common in many organizations.

Cordless phones are another example of radio frequencies being used for communication. While older cordless phones were fairly simple to monitor with a scanner capable of receiving 900 MHz, newer cordless phones often use more secure transmission methods. It is important to note that many newer cordless phones, especially those deployed in enterprises, often use encryption as well.

Keep in mind that while many of the benefits of monitoring cordless phone conversations are obvious, others are not. One great example of information that can be gathered from both wireless headsets and cordless phones is voicemail passwords. Using a DTMF decoder, you can capture the voicemail passwords entered on these wireless devices. Remember, however, that the government takes unauthorized monitoring of telephone communication very seriously, and it is essential to consult with counsel and ensure you have permission of the target and the affected staff members prior to monitoring telephone calls.

NOTE

Don't be fooled by dummy cameras. While dummy cameras often use the same housing as live cameras, they will of course not transmit any information. If you encounter a camera with an antenna, but cannot grab a signal, then assume it is a dummy camera and move on.

Closed Circuit Television (CCTV) cameras are another device that was previously tethered to a transmission cable. These days, it is common to come across CCTV cameras that transmit their signals wirelessly. Penetration testers may be used to attempting to monitor camera transmissions by penetrating the management console, however it is often possible to grab the signal from the air. This is another tool in the arsenal of the professional penetration tester. It is usually fairly simple to determine if a camera is transmitting wirelessly, as the antenna is generally obvious.

Some of the most fascinating sources of radio signals within an organization are wireless microphones. Wireless microphones are fascinating because they are often used in boardrooms, conference rooms, auditoriums, and other places where important meetings are held. This means that if you are able to intercept these transmissions, then the information is often sensitive and may be of great value. Keep in mind that of all the radio reconnaissance targets, wireless microphones will usually have the weakest signal, meaning that you will need to be in close proximity to monitor them. This means that you will often have to physically penetrate the organization before you are able to monitor the wireless microphones. One notable exception is the offsite meeting. Board meetings and other high level gatherings are often held in hotels or other quasi public places. If you are able to find out the location of a meeting whether through public media or by other reconnaissance, you may be amazed by what you hear. When in a quasi public space, you will still need to attempt to look like you belong. Depending on the scope of the engagement, if an offsite meeting is held in a hotel, it may even make sense to get a room near the meeting room.

IN THE REAL WORLD: The authors, tasked with securing a boardroom prior to a meeting about a highly secret merger, swept the room for bugs and other transmitters. Great effort and expense was taken by the organization to ensure that what was said in the room, remained in the room. About 10 min before the meeting was to begin, an administrative assistant showed up with a wireless microphone. Upon further inspection, it was noted that the transmission of this wireless microphone was not encrypted. Had this been used, anyone outside of the boardroom with a scanner would have been able to listen to everything that was said. Keep this in mind when securing your organization.

CHAPTER 4

Offsite Profiling

WHAT IS OFFSITE PROFILING?

Offsite profiling is, as the name suggests, is gathering as much information as possible prior to visiting the target site. As with many things, the more preparation that occurs prior to the project, the higher the likelihood of success. Much of the information that will be useful is public, whether intentionally or not, and the methods described in this chapter will help you find it. A few things to attempt to determine prior to going onsite include the lingo and codes used by any radio user within the target, the make, model, and type of equipment used by the target, the frequencies that they are likely to use, and whether the guard force, if there is one, is contracted or are employees of the target.

CONTENTS

What to Look For

To start, simply search the Web for the company name and some keywords related to scanner and radios. During this process, pay special attention to search results from scanner hobbyist forums. It is not unusual to find a local hobbyist has mapped out a company's radio system and even made notes about lingo, company culture, and other valuable information. Johnny Long's *Google Hacking for Penetration Testers* has many great tips and techniques that will be helpful during the process.

Press releases are often a useful resource for determining the details of a contract guard force and the radio equipment in use. Often, marketing departments do not consider security when releasing information. When a radio manufacturer or outsourced guard company wins a contract, they will want the world to know. Take a look at a few press release aggregation sites, and it becomes immediately clear that there is a great deal of valuable information

RECOMMENDED SEARCH TERMS

While searching for information on the target, enter the company name combined with one of the following keywords:

- Scanner.
- Frequency.
- Guard Frequency.
- Physical Security.
- MHz.

TIP

In the United States, the top five commercial grade two-way radio manufacturers are Motorola, Midland, Kenwood, Icom, and Vertex.

there for the taking. Announcing that large company has chosen a particular radio product is perhaps a useful marketing tool; when making a purchasing decision it is helpful to know that others have chosen the same product. If you own a smaller business, and can't afford to spend the time of money researching multiple products, then choosing the same product as a larger company may make sense. This information, however, is also valuable to the penetration tester or attacker. These press releases will often include specific information as to the type of gear purchased as well as any special features of the gear.

Aside from press release aggregation sites, there are many other ways to find these press releases. Industry magazines, Web sites, and other publications will commonly print or post press releases. Many of the major equipment manufacturers will also post press releases on their corporate Web sites. Of course, as with any other information that is sought, search engines are a great place to start. When a press release is discovered, it is essential to review the details. Be sure that the press release is not so old that new equipment has likely been purchased, and for target organizations with multiple sites, see if the press release is location specific. It won't necessarily be helpful the know the radio equipment used in the organization's west coast location if you are targeting the east coast facility.

Keep in mind that much like other equipments, a large amount of radio hardware is sold by resellers. Find out if there are any resellers near the target organization and scour their Web sites for press releases related to the target.

Social engineering, when performed in an effective manner, can also provide a wealth of useful information. Calling the target organization and posing as a

> **TIP**
>
> When posing as a salesperson for purposes of social engineering, let the target do the talking. Ask open-ended questions such as:
>
> - How many radio users does your organization have?
> - What type of special features do you use or are you considering?
> - Do you currently use encryption or trunking?
>
> The answers to these questions and others are helpful when selling radio equipment, and are also helpful when profiling a target organization.

radio salesperson can result in release of valuable information. Commonly, the facilities department will manage radio equipment. If the target has a physical security department, they may be tasked with radio procurement. Remember, before calling, you must gather as much information as possible. If you aren't able to speak intelligently about the equipment you purport to sell, the ruse will likely be discovered.

With rare exception, legally broadcasting on a frequency within the United States requires a license from the Federal Communications Commission (FCC). In addition to many other functions, the FCC ensures that the radio spectrum is used in the most efficient manner, and that multiple organizations or stations do not fight over the same frequency. It should come as no surprise that to keep everything organized, the FCC maintains a database of the licenses issued to use specific frequencies. This database includes information such as the frequency information as well as the license, including their location and contact information. What may, however, come as a surprise, is that this FCC database is available to the public (Figure 4.1). The catch is, and of course there is a catch, that the database is cumbersome and difficult to search and retrieve specific information.

The private sector has come along and created resources to ease access to the information contained in the FCC database. Radioreference.com (Figure 4.2) and nationalradiodata.com (Figure 4.3) are two such resources.

Radio Reference, while a paid site, has many free areas which contain a wealth of information. National Radio Data offers a hobbyist rate which at the time of this writing is $29 per year.

There are few things to keep in mind when searching for the frequencies in use by an organization. As discussed earlier, there are many radio equipment resellers. A large number of these resellers also provide maintenance for the equipment, and at times offer to provide frequencies to their Clients. In these cases, the frequency is licensed to the reseller and then leased to an organization.

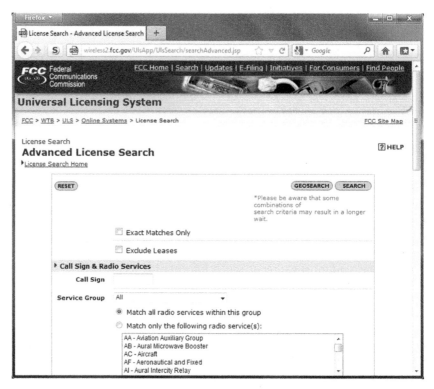

FIGURE 4.1 FCC Universal License Search (ULS) Advanced License Search Web Site

FIGURE 4.2 The Radio Reference Web Site

FIGURE 4.3 The National Radio Data Web Site

TIP

The following link can be used to access the FCC database Web site: http://wireless2.fcc.gov/UlsApp/UlsSearch/searchAdvanced.jsp. If this link does not work you can also search for "FCC ULS advanced license search" to find the site.

TIP

When searching the FCC database or similar radio frequency databases, the authors often look up not only the current company name, but also any other names under which the company may have done business. This includes subsidiaries as well as partners and acquired companies, and also the names of companies from whom the target company may have acquired facilities. For example, if the target is the operation center of "Big Orange Bank" but the operation center used to be owned by "Big Green Bank," search the sites for frequencies used by "Big Green Bank." When one company acquires another company or their facilities, they rarely replace the radio equipment at that site. Oftentimes the same radio setup used by the previous company will continue to be used by the new company.

Using RadioReference.com for Offsite Profiling

The following examples show how to use the RadioReference.com site to find information on the target. All of the following searches were performed using the free version of RadioReference.com.

For this example, we will assume you have been hired by Franklin County, in Ohio, to perform a physical penetration test of their data center. For this case Franklin County was chosen as an example as it is a county containing a large city in the state where authors live. To start off, go to frequency database on RadioReference.com, navigate to Ohio, and then Franklin County. Looking through the Franklin County section of the database you will see that Franklin County has two large trunked radio systems; one for public services and one for public safety. Studying both trunked systems will give you some interesting targets to monitor to gain information on the data center. Of specific interest are talk groups a17 and a19 on the public service trunked system (Figure 4.4). These talk groups are labeled "County Data Center Channel 1" and "County Data Center Channel 2," and are likely, based solely on their names, to provide useful information regarding the data center. The page also contains all the information needed to program this trunked system into a trunking capable scanner and monitors these two talk groups for intelligence.

RadioReference.com and the FCC database have the ability to show which transmitters are in a given geographic area. The authors often use this feature when they cannot find the frequencies used by a target organization by searching for the company name. When performing geographic based searches, the authors prefer to use the RadioReference site because they have a functional tie in with Google Maps allowing you to visually see where the FCC licenses are located on a map. This feature can be located on the Radio Reference Frequency Databases page. At the time of writing, the search was under the Search

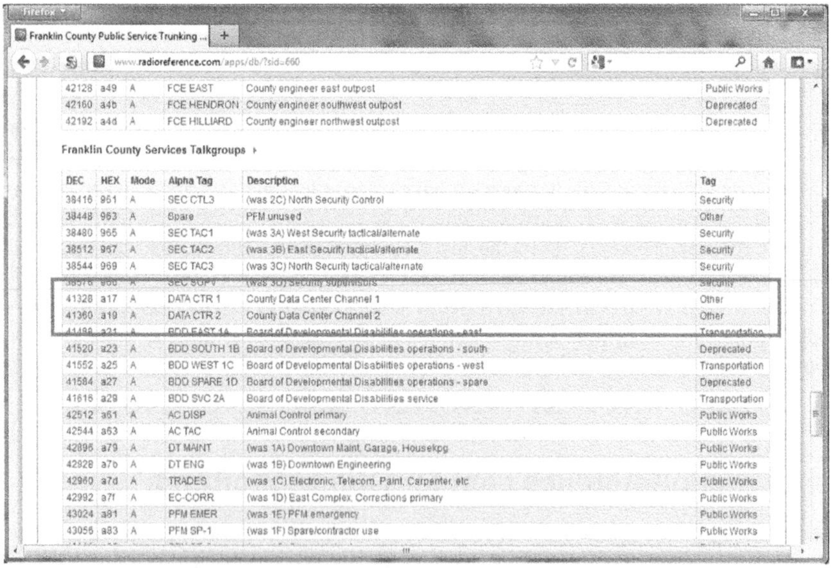

FIGURE 4.4 RadioReference.com Example

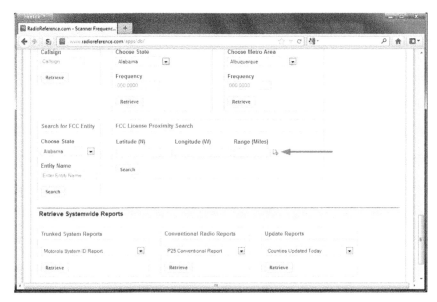

FIGURE 4.5 Launch Location Tool Helper Icon

FCC Database section of the page, with the specific search being labeled as "FCC License Proximity Search." If you know the latitude and longitude of the target organiztion, or a point near the target organization, you can enter that directly into search function using the WGS84 GPS datum format (IE, 30.29128 −97.73858). Setting the search distance to two miles is a good starting point which can then be adjusted up or down depending on the number of transmitters in the area, size of the target property, and accuracy of the latitide and longitude data at hand. If the latitude and longitude of the site is unknown, click "Launch Location Tool Helper." The link to the "Location Helper" tool icon is located next to the data entry boxes (shown in Figure 4.5) and can be a very helpful tool.

Figure 4.6 shows the results of a geographic search in downtown Cleveland, Ohio. If, using this method, a licensee on the target property is found, the name of the licensee should be used to run additional name-based searches. This will often turn up additional frequency information for remote site or licenses that are registered at a different address.

CASE STUDY: OFFSITE PROFILING

Physical penetration test at a ship dock.

The following story is, well, mostly true. Because we are professional security people, we cannot disclose details of our Clients' systems nor their names.

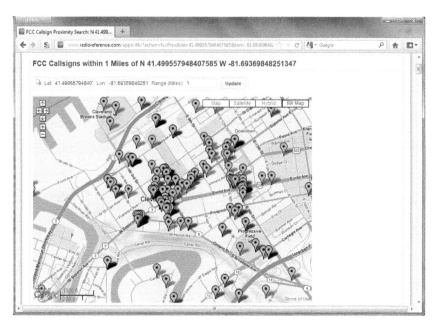

FIGURE 4.6 Example FCC License Proximity Search Results for Downtown Cleveland, Ohio

> **NOTE**
>
> The addresses in the search can be the licensee's address or the location of the antenna tower licensed to use the frequency. Always try to use a database that shows the antenna location when looking for the frequencies used at a specific site. This is especially helpful for multi-location businesses as often multiple transmitters over a 75 mile area are covered with a single, variably placed license.

All the events in this case study happened, however we drew from events from multiple penetration assessments to create this scenario.

The penetration team's goal was to gain access to the ship dock, specifically the high-security area where the ships docked, and deliver a special package to prove we were there—a loaf of bread. In addition to the usual gates around the perimeter of the property, the actual docks had additional security controls: a second fence, gates, cameras, automated ID card reader checkpoints, and guards, to prevent people from moving undetected between the ships and the mainland. We noted that the ID card readers had no controls in place to prevent tailgating, so it would be possible to sneak in if security wasn't watching the gates.

We began offsite profiling by Google searching the name of the company and common radio terms. This turned up nothing of interest. Next, we searched for the company name on Radio Reference (www.radioreference.com), and still turned up nothing. Next we navigated to the state and county inside the Radio Reference Frequency Database (Figure 4.7).

From here we opened the Businesses link under Other Agencies and reviewed the companies to see if any matched our Client (Figure 4.8). Sometimes by manually reviewing the businesses in the area, we will find a business whose name is close to or related to the target business.

After reviewing this page, the Client's name, or a related name, still did not show up. It's important to note that the Businesses listing in this section mostly includes only large business and attractions. Next we searched all the FCC licenses for businesses in the area, and again turned up nothing related to our target.

Widening our Radio Reference search to the general area turned up about two dozen shipping companies and dock names, but none of them matched the Client name. We started searching press releases to see if there was any evidence of a relationship between any of the companies we found and our Client. When we look for press release information, our first tool is Google, but we also make use of the Client's Web site, and any affiliated companies we know they work with. If we know of any integrators in the area, we also search their Web sites for press releases about business dealings they may have with the Client company.

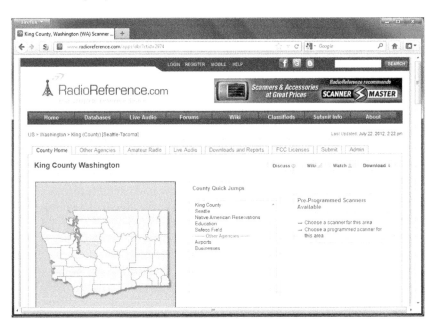

FIGURE 4.7 Radio Reference Frequency Database Page for King County Washington

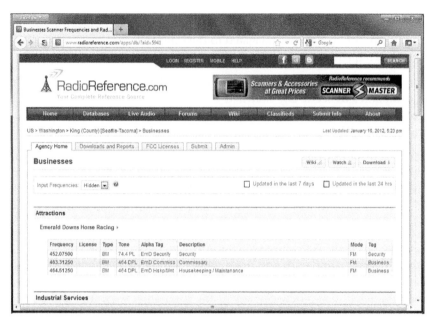

FIGURE 4.8 Radio Reference Frequency Database Businesses Listings

For this Client, these basic searches did not turn up anything. This was very odd, because large sites with lots of logistical needs invariably employ a radio system to help coordinate activities.

Unable to identify a relationship between the Client and the companies in the area, we used the map feature on Radio Reference. We used this feature to find transmitters in the area, and located a transmitter tower on the Client site. But the licensee did not match the name of our Client. Doing further research turned up information that established the connection we were looking for: the FCC license was still in the original name of the company, which had changed as the result of a buyout. The parent company, our Client, had never bothered to change the name on the license. This is not an uncommon situation to run into.

Once we had established the connections, we now had useful information that we could act on. Thanks to Radio Reference, we knew they had a conventional four-channel radio system. In the Radio Reference database, Channel 4 was labeled "security." The other three channels were used for operations.

Armed with this information, we arrived onsite and programmed all four channels into our scanner and started listening. Before long, we were able to confirm that Channel 4 was in use by security. We monitored this channel. Radio chatter confirmed that there were guards stationed at the main gate, but we never heard any transmissions from the second gate.

SEARCHING FOR FCC LICENSES FOR BUSINESSES

To search all the FCC licenses for businesses in the area, it's necessary to navigate back to the County level in the database and open the FCC Licensees tab. From here select Display All FCC Licenses for XXX County. This will open up a page displaying all FCC licenses in the area. By default this view only shows the public safety licenses in the area. To display the businesses in the county turn on the Business Conv and Business Trunked filters at the top of the page. If you want to only see the businesses, turn disable showing the public safety entries by turning off the Public Safety Conv and Public Safety Trunked filters (Figure 4.9). This will usually create a very long list. If you have a subscription to Radio Reference you can download this data to Excel using the Downloads and Reports tab. Next you can use Excel, of other spreadsheet or database tools, to search the information. In the downloaded file, GB, GJ, GS, GO, GU, GX, IG, IK, LN, LR, QT, YB, YG, YI, YJ, YK, YS, and YX are the tags used for Business licenses. By far analyzing this data offline is the recommended approach to take. If you do not have a subscription, you will need to search the information online. Licenses are listed alphabetically, start by navigating to the page where the company name should appear to see if a close name can be found.

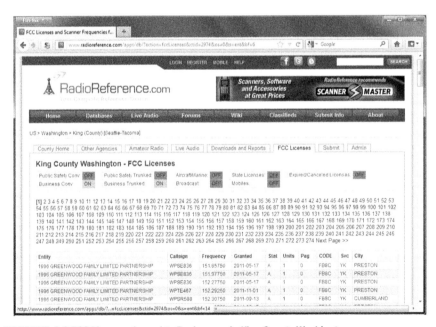

FIGURE 4.9 FCC Licenses Issued to Businesses in King County Washington

Using Google Maps, we got an aerial view of the property, which showed two roads leading into the property. Both roads were gated at the perimeter. One clearly had a guard booth; the other, we couldn't tell.

> **TIP**
>
> If you do not find the target listed in the FCC database, you can also set up the database to display expired FCC licenses by turning on the Expired/Canceled Licenses filter. The authors have encountered a couple of occasions where companies are still operating on expired licenses. If this is discovered, be sure to notify the Client because they are illegally transmitting on those frequencies.

Security appeared to have one mobile unit, which used an SUV to cover the grounds. This unit did rounds when ships were not at dock. However, when a ship was at the dock, we were able to infer from the radio chatter that it parked and watched the "high-security gate" leading to the dock. Additionally, we heard two or three guards on foot, depending on the time of day.

From the map, we knew that the property also had a number of train tracks leading into it. From what we could tell, the rail yard area was patrolled by the mobile unit when it was doing rounds. So, when a ship was docked, the train track entrance was not patrolled. Furthermore, while we were at the site, we could see that the train tracks were not gated. Our reconnaissance found that the mobile unit patrolled this area as part of its rounds. This meant that when a ship was docked, the train tracks were not patrolled. And although the dockyard had a fence around it, most of the dockyard was open via the train entrance, which did not have a fence or gate.

While doing onsite reconnaissance, we saw security cameras, and the area was lit with flood lights and street lights, but coverage appeared to be mainly focused at the gates and water. On our map, we noted the areas that had camera coverage, as well as blind spots, for later debriefing with our Client. After the test, we would provide the Client with recommendations on where to place additional cameras to fill in these gaps.

By monitoring the other channels, we could also hear the operations side of the dock and learn when ships were coming and going. We could also learn when ship crew where leaving for shore leave, and when they were supposed to report back. All of these activities were coordinated, so usually we started hearing radio chatter well before the ship crews would go ashore. By monitoring the guards at the perimeter gate, we could learn when ship crews were coming back. Once they passed through the perimeter, the gate guards would notify the mobile unit of the inbound arrivals heading to the docks, so that there would be someone at the inner gate to get them through.

We also observed the type of equipment and outfits being worn by the workers and dock hands so we could get similar gear. The company we were working for is very concerned about safety, so before we arrived onsite we were required to purchase extra safety gear and required to wear it while on property. Coincidently, we had been required to wear this same type of safety gear while doing a physical penetration test at another site so it was already dirty and broken in. This helped us to blend in. We would have purchased similar looking gear locally had we needed.

Armed with this information, we were able to plan an approach to gain access to the facility. We waited for a ship to come into port. By listening to their radio chatter, we could hear them preparing for the arrival of a ship several hours in advance. Initial chatter would begin to happen about 12–24 h before a ship's arrival in port. Chatter on the operations channels increased as the time of arrival drew near. Had we needed, we could have learned everything about their procedures for loading and unloading this way, as well as kept tabs on where they were in the process. Anticipating the ship's arrival, the guard in the mobile unit had a habit of parking in a position where they could observe the inner gate a short time before the ship docked. When in this position, they were not able to monitor the perimeter at the train tracks. This was the time when we would be able to penetrate the perimeter by following the rail lines. Once inside the perimeter, we found a hiding spot where we could watch traffic going in and out of the higher security area, and waited for an opportunity to enter ourselves.

Given the attention on this area we didn't feel a technical attack (lock picking, fencing climbing, etc.) would be the best way to go. So we decided to try to use a stealth approach, augmented with social engineering (our costumes and equipment, to allow us to blend in and look like employees, and knowledge of their processes and terminology that we had gained through radio surveillance) in case we were detected or needed to reveal ourselves. We had heard on the operations channel that they were going to bring the cargo onto the ship in an hour. From listening, we learned the cargo was a mix of shipping containers and smaller items inside semi-trailers.

We went to the cargo area, and found an unlocked trailer. This part of the property didn't seem to have guard coverage at the moment. We entered through a side door in the trailer, which was fortuitous because that door happened to have an interior handle which allowed us to open the door from the inside. The rear cargo doors did not appear to have a way to unlock them from the inside.

We hid out in the cargo area and waited. Around 90 min later, they started moving the cargo.

We could verify this by the radio traffic (using headphones to keep our noise down) and noises heard in the cargo waiting area. After a while, our truck started up. It drove for about 5 min, then stopped. Once it stopped we waited several minutes until it was quiet. Once we felt safe, we carefully opened the side door, looked around to make sure no one was in the area, and exited the trailer. At this point, we proceeded to obtain evidence that we were inside the target area by taking pictures. We also found a place to leave the loaf of bread.

Having accomplished our objective, we exited the area. This was accomplished by simply walking out the gate. The guards assumed that we were dock hands, and did not check our IDs. We learned later that they were supposed to perform an ID check, but had fallen out of this habit when the ship crew wasn't present and only "trusted" dock hands were present. With the number of personnel employed at the dock, and turnover in the workforce, it was easy to pass by without raising suspicion.

Remediation and Lessons Learned

As a result of our successful penetration, the dock company became far more diligent about having people scan ID cards to get in and out of the secure area. The benefit that really sold them here is a matter of life safety in an emergency response situation. Knowing how many people are in an area, in the event an accident occurred is very beneficial information. If you send first responders into a fire, it's great if you can tell them we have X number of people unaccounted for, instead of "a few" people.

We recommended the Client install turn styles or other anti-tail gating devices to prevent multiple persons from entering or exiting on one card swipe.

The Client made changes so the mobile patrol remained mobile even when ships where in dock.

The Client set up a video system that could monitor the secure dock entrance, so the mobile unit was no longer required to monitor that gate.

Additionally, we recommended the Client install a smart video system to watch the train tracks and automatically notify the guards if the system sees people. This was also a life safety issue. In the past, there had been incidents with unauthorized persons wondering onto the tracks, and getting into the facility. While preventing their entry into the facility was obviously desirable, it was also important to remove them from the train tracks as fast as possible to avoid being hit by a train.

Lessons:

- *Perimeter security.* If there is an area that is impractical to fence and gate, it diminishes the security provided by whatever perimeter fencing you do have. Unsecured perimeter areas should be subject to increased monitoring.

- *Cameras and lighting.* We noted on our map the locations and coverage areas of all the cameras and lights, and made recommendations to the Client to cover blind areas, particularly in the vicinity of the train yard, where the exposed perimeter allowed us to enter the site undetected.
- *Lax security posture.* Security staffs need to establish and follow procedures closely, particularly at gates and other checkpoints. This is a constant battle to keep people following process. Without external reinforcement of process, it is simply human nature for people to relax as the job becomes routine, and this leads to taking shortcuts. First, we recommend to automate process where possible, because machines are much more reliable than people when it comes to following a process exactly the same way every time. When a person is required to follow a process repeatedly, it becomes routine, and soon the person stops thinking. We find it preferable for the human elements in the security chain to think, and so we recommend reserving them for tasks that require thinking and judgment, which are domains currently still better suited to human beings. To the extent that people do need to adhere to process, spot checks and audits are helpful. For Clients that are concerned about this, we help them establish re-occurring physical penetration tests every quarter or so to test the security force. If they know they are being tested they tend to maintain a more ready posture. Finally, we recommend regular training and drills. When we run into cases of a laxed guard force the first thing we recommend is more training, and not firing the guards. Firing the guards and bringing in new ones who will also not be trained doesn't fix the problem.
- *Inadequate guard coverage.* The mobile unit performing double duty at the inner dock gate was a cost-saving measure that left the Client vulnerable. When the mobile unit needed to be at the gate, it left the train yard unguarded. Better coordination with the foot patrol potentially could have helped remediate this hole. Additional lighting and cameras would have provided additional layers of security, further strengthening the perimeter hole. We recommended a smart video system, which can detect humans over other items automatically, which eliminates the need for a human to monitor the cameras. Even the best trained person will tune out after watching a video screen for a while. This frees the human to perform more valuable duties. Wherever possible, we like to recommend replacing humans performing non-thinking tasks with technology that is more reliable, while allowing that person to do something more rewarding.

CHAPTER 5

Onsite Radio Profiling

Once we have gathered as much information as possible offsite, it is time to go onsite and perform additional reconnaissance. Remember, of course, that the techniques described here are for profiling a target for which you have explicit authorization. Hearing the chirp of a police radio while performing a penetration assessment is rarely, if ever, positive.

The goal of onsite profiling is to discover the frequencies used by the target organization, and to validate the frequency information obtained during off-site profiling. We are looking for the frequencies used by the guard radios, facilities radios used by janitors and maintenance staff, wireless headsets if in scope, and any other radio traffic that may emanate from the target organization. This profiling does not require a special trip; it should be included in the normal onsite profiling that is part of any physical penetration test.

INITIAL ONSITE RECONNAISSANCE

With any reconnaissance, it is essential to have an idea of what to look for. There are several things that you should keep an eye out for on any onsite reconnaissance mission. Be sure to determine the number of guards if possible. This may require some inference, as for every guard that is visible from the perimeter, there are likely several more in other locations. Once the frequencies used by the guard radios are discovered monitoring the guards' radio traffic is another way to verify the number of guards onsite. Most online map services offer aerial views which are a great starting point when you need a sketch of the target site. A sketch of the target facility will be helpful to mark locations of guards, antennas, cameras, and any other useful information. Pay attention to any antenna seen on the property or attached to the building. Note the type of antenna and size to try to determine the frequency range it operates on. Also look for cameras,

> ## TIP
>
> Different online mapping sites often have different sources for their aerial photography. Thus, when generating a map of a site, it is often beneficial to check the aerial photography from multiple online mapping services such as Google Maps and Bing Maps. For example, Bing Maps aerial view and bird eye views often contain images from different sources, which can allow you to view the site at different times of year, or from different angles.

and see if they appear to be wireless, most wireless cameras have an antenna on them or are attached to a transmitter box with an antenna coming out of it.

THE GUARD FORCE

If there are security guards, it is essential to learn as much about them as possible while profiling. This can easily mean the difference between a successful and an unsuccessful penetration test. Security Guard uniforms hold a great deal of information. Contracted guards will often wear the uniform of the contracting company, and knowing the name of the guard company allows you to use the techniques in the offsite profiling chapter to search for the frequencies they use. It is also common that there will be a mix of contracted guards and guards employed by the target organization. This is common in both the government and private sectors, and it is likely that in this situation the contracted and employee guards will be dressed differently. The contracted guards will generally wear standard guard uniforms, and the employee guards will often wear ties and blazers or suits. The employee guards are usually in locations where they interact with other employees, such as at the reception desk or assigned to executive protection while the contract guards will generally be found patrolling the perimeter or guarding remote locations. If contract guards are present, it can take a week or more of observation to determine their shift rotation. Also note if the guards are the same each day. While it is uncommon,

> ## TIME IS OF THE ESSENCE
>
> ### Start a Frequency Counter the Minute You Arrive Onsite
> Time is usually of the essence during the onsite profiling phase, as the longer you are onsite, the more likely you are to be discovered. The more tasks that you can run concurrently the more efficient your reconnaissance will be. We recommend setting up the Scout frequency counter (Figure 5.1) (described in detail in this chapter) upon or prior to arrival, and letting it run while you are performing visual reconnaissance. The longer the Scout is able to run, the more information it will gather.

as a company is usually assigned a specific set of contract guards, if there is high turnover with the contractors it will likely make it easier to impersonate a guard and gain access. Discovering guard names, whether from nametags, badges, or desk placards can also prove valuable. Remember, the more you know about the target environment, the better the chance of a successful penetration assessment.

USING A FREQUENCY COUNTER

A portable frequency counter (Figure 5.2) is one of the most valuable pieces of equipment for onsite profiling. They work by counting the number of beats, or oscillations in the airwaves of a set period of time, giving the frequency of those oscillations and thus the transmission frequency. An important caveat is that if there are multiple transmission frequencies in the vicinity, the frequency counter will give false data as it will likely combine the oscillations and give an inaccurate frequency count. Frequency counters work best at counting a single signal in a low noise environment, as they need a signal that is high enough over the noise floor that the counter can separate that signal from the background noise and only count that waveform. The frequency counter will display the closest "near field" signal, which must be approximately 10–15 times as strong as other signals in the area, and the closer the frequency counter is physically to the signal, the better the chance that the frequency counter will be effective. It is becoming more and more difficult to find a place where there is single signal in a low noise area, as today's world is an extremely RF-rich environment. With the proliferation of wireless phones and the wireless network towers, it can be a challenge for the frequency counter to lock onto a signal. This is especially difficult when in a signal rich urban area or when there are multiple transmitters in a single tower. Additionally, commercial radio, pager, television, and emergency dispatch towers can cause havoc for the frequency counter. This makes sense, as radio, television, and emergency

FIGURE 5.1 Optoelectronic Scout, Reprinted with Permission from Meagan Call

FIGURE 5.2 Optoelectronics Cub Portable Frequency Counter, Reprinted with Permission from Optoelectronics, Inc.

TIP

WiNRADiO, Optoelectronics, and others make filters to remove common strong frequency ranges such as the commercial FM broadcast band and page broadcast bands. These can be put inline between the antenna and a frequency counter to filter out these strong signals. The WiNRADiO WR-UBF-1800 filters can be configured to block broadcast AM as well as broadcast FM signals. The Optoelectronics N100 FM Notch Filter (Figure 5.3) blocks the FM broadcast band and is designed to fit inline between the antenna and the frequency counter and requires not additional cables. This makes the N100 ideal for mobile and handheld operations. Unfortunately for the penetration tester, TV stations cover such a wide range of frequencies they do not offer a filter to remove those strong signals. However, a tunable notch filter (a filter used to eliminate limited ranges of frequencies based around a central frequency) can also be used to prevent a single strong frequency from interfering with a frequency counter.

FIGURE 5.3 Optoelectronics N100 FM Notch Filter, Reprinted with Permission from Optoelectronics, Inc.

TIP

Sweeping for bugs? A common trick used when sweeping for bugs is to take a footprint of the frequencies that are present in an a room or area about 100 ft from the area to be swept for bugs, and then take a footprint of the frequencies in the area to be swept for bugs. Anything that is actually transmitting in the area you are sweeping, even if transmitting at very low power, will appear much stronger due to the law of inverse squares.

dispatch towers must have the power to broadcast over a geographically large area, while the two-way radio transmitters used by many organizations must only blanket a comparatively small area. The law of inverse squares means that the closer you are to a signal, the stronger the signal will be. It should, of course, be obvious that a handheld frequency counter will generally allow you to get fairly close to the signal source. This is especially true when you have access to the interior of the facility that is being targeted.

For penetration testers we recommend an Optoelectronic Scout frequency counter (Figure 5.4). The Scout has many advantages over a normal frequency counter, and the extra features are very useful for a penetration tester. The Scout can hold multiple frequencies as well as multiple hits per frequency, and the frequencies can be downloaded to a computer for easy analysis. This can

FIGURE 5.4 Optoelectronics Analog Scout, Reprinted with Permission from Optoelectronics, Inc.

be a huge time-saver, as reading frequencies from a frequency counter screen and entering them into a computer can be a real headache. The Scout also offers reaction tuning, which means that it can automatically tune a compatible scanner to a frequency seen by the Scout. While this has limited usefulness, we have found that in some situations it can be helpful and save time. For example, if once onsite there is not enough time for proper profiling, and reaction tuning may allow you to quickly monitor signals in the area. While this is no substitute for proper profiling, it is a good hack if needed. Additionally, if the budget allows it, an extra scanner doing reaction tuning during an assessment can be a great way to catch frequencies missing during the onsite and offsite profiling.

What is most useful as far as the Scout, however, is its small size and portability. The Scout can be kept in a backpack or worn on a belt under a coat, which is helpful in situations where discretion is necessary. Being seen removing and replacing a frequency counter from a backpack or coat pocket is suspicious and can draw unwanted attention during a penetration assessment.

The Optoelectronic Scout comes in both analog (Figure 5.4) and digital (Figure 5.5) versions. The digital version, as you may suspect, is designed to find the frequencies of a radio operating in a digital mode, which we have found to be of limited use during penetration assessments. Additionally, the digital mode tends to be finicky outside of a lab environment. The analog version, however, is a great piece of equipment and we highly recommend it.

If you choose to get a frequency counter other than the Scout, be sure to get one with a hold function which will continue to display the frequency after the counter is out of range of the frequency. The hold function will save in its memory the first frequency the counter locks onto, which saves you from having to write down the frequency with pen and paper. Of course, there is a caveat that even with a hold feature, the benefit is limited as there is generally a good chance that there will be a lot of trial and error as the frequency counter will lock onto the first strong signal it detects in the area which may not be the signal you are ultimately looking for. This is another reason to

---

TIP

People tend to want to help, it is human nature. Combine this with the fact that most people love to talk about their jobs, and it can be simple to capture a frequency. There have been times when we simply started a conversation with a guard about their radio, expressed an interest in radio technology, and asked the guard to see their radio. From there, it is simply a matter of using a single memory frequency counter to grab the target signal.

FIGURE 5.5 Optoelectronics Digital Scout, Reprinted with Permission from Optoelectronics, Inc.

use a Scout, as it will also capture unwanted signals, however it will hopefully also capture the signal that you are looking for. All models of the Scout also include reaction tuning. This means that if you connect the Scout to a compatible scanner (a list of compatible scanners is available on the optoelectronics web site), the scanner will tune to the frequencies that the Scout counted.

If you are new to radio reconnaissance, and don't want to make the investment in a new Scout, a used one may be a viable option. There are, however, issues with buying any used piece of equipment in general, and issues specific to buying a used Scout. It is almost certain that in a well-used Scout that the battery will be beyond its prime and likely close to the end of its useful life. If you

> **NOTE**
>
> It should go without saying that since it requires soldering, replacing the battery on your own will void whatever warranty you may have received with your equipment. Additionally, you make any changes to your electronic equipment at your own risk; the authors cannot accept responsibility should any repair or replacement damage your device.

are comfortable working with electronic devices, and soldering in particular, replacing the battery can be a fairly simple job. Hackers tend to be tinkerers at heart, so when replacing the battery, it may make sense to install a replacement battery that will hold its charge longer than the OEM battery. If doing this upgrade, make sure that the replacement battery is compatible with the Scout's charging circuit.

Another common issue with a used Scout is a blown front end, the initial receive circuitry just after the antenna, an essential part of the receive portion of any radio. This happens when the Scout has been exposed to an unusually strong signal. Optoelectronics offers repairs, however at the time of this writing the minimum repair fee is $100, which is about one third of the cost of a brand new Scout.

Certain models of scanners have frequency finding features. Band scope is one such feature, however it has limited usefulness in reconnaissance unless it has a fast scan rate. Additional information is available in Chapter 7. Frequency grabbing features on scanners can be useful, and on higher end equipment there is often an option to automatically store detected frequencies. Keep in mind that the frequency grabbing features on these scanners function similarly to other frequency counters, and have short-range capabilities and lock onto strong signals. Multiple manufactures use different trade names for this capability, and a few of them are described in the sidebar.

When using a signal counter, or the signal counting capabilities of a scanner, there are several tips that will make success more likely. Although it may seem counter intuitive, the signal strength does not always get stronger as you move toward the transmitting antennas. In fact, the base of the antenna tower is one

> **NOTE**
>
> Scanner manufacturers use several names to describe their frequency counting capabilities:
>
> - Uniden Close Call™.
> - Radio Shack Signal Stalker I and Signal Stalker II™.
> - GRE Spectrum Sweeper™.

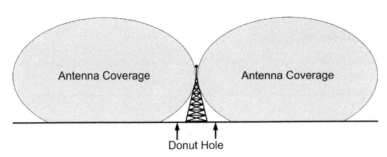

FIGURE 5.6 Donut Hole Created Close to Transmitter Towers

of the worst places to measure a signal, as radio waves form a donut shape when leaving omni-directional antenna, with the antenna in the middle of the donut hole (Figure 5.6). This center of the donut hole is essentially leaving a "cone of silence" around the tower.

VISUAL RECON

As with other types of reconnaissance, a sharp eye is an essential tool in radio profiling. Being able to examine a radio used by the organization, whether from a distance or up close, can provide a wealth of information. Often, organizations will place stickers on the radios listing the channel assignments, for example channel one is security while channel 2 is maintenance. It is also common that these channel listings will be posted in guard stations or behind the reception desk, so be sure to look around.

Identifying the model of radios in use by the target organization is invaluable. Many radios have the manufacturer name prominently displayed on the outer housing, and some will even have the model or series number. This is another place where having a camera can be helpful. A quick snapshot will allow you to take all the time you need to research and determine the radio make and model once you are offsite.

Once you have determined the type of radio in use by the target organization, there are a wealth of resources to enable you to gather a large amount of

---

TIP

There are many reasons to have a camera with a zoom lens with you during a penetration test, and taking pictures of equipment or frequency lists is one of them. If your frequency counter does not have a memory function you could also use the camera to take pictures of the frequencies displayed on your frequency counter for later review.

TIP

In the United States, the top five commercial grade two-way radio manufacturers are Motorola, Midland, Kenwood, Icom, and Vertex.

NOTE

It should go without saying that we do not encourage nor condone forming relationships with people in the radio community solely to gather information.

Once you have determined the model, or models, of radio in use by the target organization, it is time to get back to research. Knowing the model of the equipment will let you learn what frequency ranges are supported by the radios, and if it has any special features such as encryption, trunking, or is digital.

information. Internet search engines and web sites (Figure 5.7) are an obvious choice, however there are also people who will likely be willing to help you. Developing a positive relationship with local radio equipment dealers and

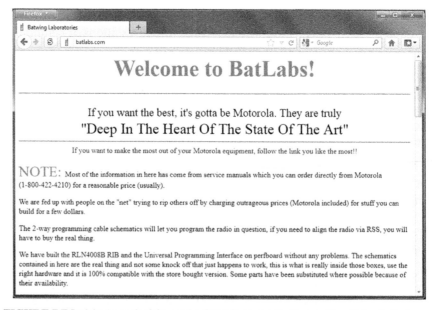

FIGURE 5.7 Batlabs (www.batlabs.com) a Great Resource for Researching Motorola Radios, Although it is Not Updated as Often as it Used to Be

Ham radio operators will provide experienced input. We may be biased, however we have found that people in the radio community tend to be friendly and generous with their time and knowledge. As you gain experience with radios, you will be able to quickly determine at least the manufacturer of a device with a quick glance.

Antennas

A lot can be learned from the type of antennas on radio equipment. When looking at the radios used by the target organization, be sure to note the antennas length and type. Antennas types are described in greater detail in the Antenna section of Chapter 2, however it is good to know that as a general rule the shorter the antennas, the higher the frequency (such as 460 MHz) and the longer the antennas the lower the frequency (such as 150 MHz) used by the radio (Figure 5.8). The main exceptions to this rule are compact antennas which are designed to be small in size, and still operate on lower frequency ranges. These antennas may be ½ or ¼ (or even smaller) wave antennas.

Like everything else in this book, and penetration testing in general, the more experience you have the easier the identification of antennas and their probable frequency ranges will be. Having a properly tuned antenna is far more important when transmitting, in fact, transmitting with an improperly tuned antenna can damage the transmitter. This is another place where being able to take a picture of the target radio is helpful, as it will allow for you to research the antenna at your convenience. Manufacturer web sites and catalogs will

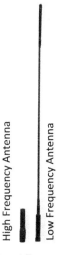

FIGURE 5.8 Two Antennas Used for Different Frequencies, Reprinted with Permission from Meagan Call

often provide a description of the antenna, and in the best case, will provide the frequencies on which the antenna is designed to operate.

While onsite, also look to see if all the radios in use are the same model, or if there is a mix of types. If there is a mix, then it is unlikely that the target organization is using any special or advanced features such as encryption. The nature of radio equipment makes it such that it is difficult, and often impossible, to get advanced features to work between equipment from multiple manufacturers. In fact, it can be complicated to get features such as digital transmission and encryption to work reliably between different model equipment by the same manufacturer. Additionally, if there is a mix of equipment manufacturers, it is unlikely that trunking is being used.

SEARCH COMMON FREQUENCY RANGES

Searching common frequency ranges is labor intensive and time consuming, however if signal counting and the methods described in the {cross ref offsite profiling chapter, with page number} prove unsuccessful, this may be the only option. The concept is fairly simple, listen on commonly used frequencies and look for traffic from the target. The good news is that if you are able to view the radios in use, you can generally narrow down the number of possible frequencies and save significant time. The following describes the common frequencies in use.

Family Radio Service (FRS)

FRS was originally designed to provide a low cost option for families to use walkie-talkies. The idea of families using walkie-talkies may seem quaint now with the proliferation of wireless phones, however it was of course conceived prior to the cellular age. FRS is now commonly used by businesses to provide a cheap radio communication solution. Power is limited to a half watt (.5 W).

Family Radio Service Frequencies (*Note:* These are United States Frequencies)

Channel	Frequency (All frequencies are in MHz)
1	462.5625[a]
2	462.5875[a]
3	462.6125[a]
4	462.6375[a]
5	462.6625[a]
6	462.6875[a]
[a] Shared with GMRS.	

Family Radio Service Frequencies (*Note:* These are United States Frequencies)

Channel	Frequency (All frequencies are in MHz)
7	462.7125[a]
8	467.5625
9	467.5875
10	467.6125
11	467.6375
12	467.6625
13	467.6875
14	467.7125
[a] *Shared with GMRS.*	

General Mobile Radio Service (GMRS)

GMRS is less common than FRS as it requires a license to use the service. The license is not challenging to obtain, as there is no required exam and at the time of this writing the cost is $85 and a single license can be used by an entire family. A main advantage is that GMRS allows the use of repeaters and more transmission power than FRS.

The names of the channels in GMRS often vary by manufacturer, however the frequency is what is important for our purposes so they can be programmed into a scanner to determine whether the target is broadcasting on them. As with other frequencies, if possible, program these frequencies into the scanner prior to going onsite. The 8 GMRS frequencies in the GMRS Frequencies table are in addition to the seven in the FRS table listed as being shared with GMRS.

GMRS Frequencies

Common Channel	Frequency (MHz)	Note
15	462.5500	Simplex and repeater output
16	462.5750	Simplex and repeater output
17	462.6000	Simplex and repeater output
18	462.6250	Simplex and repeater output
19	462.6500	Simplex and repeater output
20	462.6750	Simplex and repeater output
21	462.7000	Simplex and repeater output
22	462.7250	Simplex and repeater output

Generally, it is only necessary to monitor simplex or repeater input; if you are able to hear the input you will be able to hear the output. Conversely, if you are able to hear the input you may not be able to hear the output.

GMRS Repeater Inputs

467.550 MHz
467.575 MHz
467.600 MHz
467.625 MHz
467.650 MHz
467.675 MHz
467.700 MHz
467.725 MHz

Multi Use Radio Service (MURS)

MURS is an unlicensed radio service commonly used by businesses.

MURS Frequencies

Channel	Frequency (MHz)
1	151.820
2	151.880
3	151.940
4 (AKA Blue Dot)	154.570
5 (AKA Green Dot)	154.600

Dot Frequencies

Dot frequencies are common business band frequencies.

Dot Frequencies

Color Dot	Frequency (MHz)
Red dot	151.625
Purple dot	151.955
Blue dot	154.570
Green dot	154.600
White dot	462.575[a]
Black dot	462.625[a]
Orange dot	462.675[a]
Brown dot	464.500
Yellow dot	464.550
J dot	467.7625
K dot	467.8125
Silver star	467.850
[a] Also used in GMRS.	

Dot Frequencies

Color Dot	Frequency (MHz)
Gold star	467.875
Red star	467.900
Blue star	467.925
[a] Also used in GMRS.	

COMMON RANGES

Not every target will broadcast on a common frequency, however they will generally broadcast in a common range. Some scanner models will have search bands preprogrammed for businesses, police, fire, and other channels, which may be useful. If you are using a scanner which supports custom band scans, enter the common ranges prior to going onsite so you can quickly scan the common ranges. With any penetration test, the more work you can do offsite, the quicker the onsite portion, and the lower the likelihood of being detected. Manually searching ranges can be a very time-intensive process. You can often narrow down the ranges to be searched by using information gathered earlier in the onsite profiling such as the types of radios used and antennas on the radios. For example, if the radios used by the guard force only support frequencies between 450 MHz and 470 MHz that would be the range you would want to search first.

Common Business Ranges

Common Business Ranges

150–174 MHz
420–425 MHz
450–470 MHz
851–866 MHz

NOTE

Where did the term dot frequency originate? While no one can give a definitive answer, the often-repeated rumor in the radio community is that manufacturers placed round colored stickers on radio equipment to make it easier to purchase multiple devices that operate on the same frequency.

Common Cordless Phone and Headset Ranges

Cordless phone and headsets commonly operate on the ranges shown in the cordless phone and headset ranges chart. It is especially important to ensure that profiling these communication devices is within scope and legal. It is rare to find equipment operating in the 43.7–50 MHz range, however the authors have seen it in use in a datacenter cordless phone during a previous penetration assessment.

Cordless Phone and Headset Ranges

43.7–50 MHz
902–928 MHz
2400–2483.5 MHz (usually digital)

SCANNER TIPS

Prior to going onsite, it is helpful to enter the FRS, GMRS, MURS, Dot frequencies, and cordless phone and headset ranges into banks on your scanner to make it easier to scan the channels. If the scanner supports banks of multiple sizes, each service can be given its own bank. When a frequency range is common, we will often put them in the higher banks, and use the lower banks for frequencies specific to the target.

FINDING TRUNKED SYSTEMS

Trunked systems present a unique set of challenges, however they are generally only used by Government entities (Local, State, and Federal), Utilities, educational institutions, and other large organizations and corporations. In larger cities, we will also likely encounter companies that lease a talk group on a trunk system run by a third party in the area. In the best case scenario, we discovered the trunking information during the offsite reconnaissance, as manually finding and mapping trunked systems can be extremely time consuming and very challenging. This, however, does not mean that it is not possible. Some trunked systems will only require you find the control frequency to be able to monitor traffic, while other systems will require you to map out all frequencies in the system. If you are using a scanner that supports Control Channel Only trunking and are targeting a compatible trunked system, you will only need to program the control channel into the scanner to monitor the trunked system. Unfortunately for the penetration tester, only select Motorola trunked systems support this.

Regardless of the type of trunked system you are targeting, it will be necessary to come up with a talk group mapping so you can separate the different

> **TIP**
>
> It is worthwhile to try Control Channel Only (CCO) first. If this works, you have saved yourself a great deal of time and effort. If it doesn't work, you have only lost a minute.

channels. The talk group mapping will allow you to determine which talk groups are used by, for example, the maintenance department and which are used by security. While time consuming, talk group mapping is not complicated, and depending on the chatter on the frequencies, it can be interesting.

CASE STUDY: ONSITE PROFILING

Internal penetration test at a large insurance company located in a downtown of a large city.

The following story is, well, mostly true. Because we are professional security people, we cannot disclose details of our clients' systems nor their names. All the events in this case study happened, however we drew from events from multiple penetration assessments to create this scenario.

Initially, our scope included only the internal network, we were looking at attack vectors coming from someone who had already gained physical access to the building, or a rogue employee. During a walkthrough of the building, we noticed that employees on the phone were using wireless headsets (Figure 5.9). People often assume wireless headsets have a very short range and can't be monitored outside the building. We talked to the Client to see if we could expand the penetration test to include monitoring the headsets. The Client agreed to include the headsets, so we worked on changing the rules of engagement to include them. For this, we needed to engage our lawyer, the Client's legal staff, and Human Resources, to make sure this was done in a way that did not violate any laws or internal company policies.

We started the penetration test by gathering information. Using Yaesu VX-5r (Figure 5.10) and Uniden Bearcat BCD396XT (Figure 5.11) scanners we search for signals in the frequency ranges commonly used by wireless headsets, reference the Common Cordless Phone and Headset Ranges section earlier in this chapter.

Our scans detected a number of signals of interest in the 900 MHz range. After completing the initial survey to see what signals we could detect, we needed to verify which ones were inside the target building. This was an important step

FIGURE 5.9 Wireless Headsets Observed Onsite

for legal liability, because we needed to ensure we did not monitor any headset, and therefore phone calls, from other business in the area.

To accomplish this, we switched to directional antennas, and, using them from several locations, determined which signals originated from inside the target building. Once we had identified which signals originated from within the Client site, we started monitoring these signals.

FIGURE 5.10 Yessu VX-5R Amateur Radio, Reprinted with Permission from Meagan Call

This lead to our first critical discovery. It turned out that one of the groups that used wireless headsets was the helpdesk. The helpdesk group handles password resets. By monitoring their headsets, we could listen in on password reset conversations taking place. Once a password reset took place it was a race to login with the new password before the user did. If we won the race, we were in and had access to whatever that user account has permissions to.

Once we logged in, we did need to set a new password. Meanwhile, when the legitimate user tried to reset their password using the temporary password provided to them by the helpdesk, they would find that the temporary password

FIGURE 5.11 Bearcat BCD396XT Scanner,
Reprinted with Permission from Meagan Call

given to them over the phone by the helpdesk didn't work, and would need to call back to have it reset again. This could have clued someone in that something wasn't right, but it's human nature to assume a simpler, less sinister problem—such as the user heard the password incorrectly, or simply to blame the computer. It's not uncommon for a password to need to be reset a couple of time even when things are working correctly. So this behavior didn't raise any suspicions.

Once we logged in and got a valid Windows authentication token, it didn't matter if the password was reset again. Windows authentication tokens are good for a number of hours. So we could use the login as a beach head to attack the rest of the network. Next we used local privilege escalation exploits to gain administrative access over the system. Once we had administrative access to the system we searched for the token of a domain administrator who had recently logged into the system and used them to gain control of the corporate domain.

If we could gain access to a list of user accounts, somehow, such as by sniffing network traffic, or finding a hard copy posted on a bulletin board or left on a desk, we could have targeted a specific user account by attempting a number of failed login until the account became locked, and needed to be reset by the helpdesk. Normally a locked account does not need to have the password reset, but if we continued sending bad authentication attempts so that the account keeps getting locked before the user can get in, eventually the user will request a password reset. So instead of waiting around and hoping to hear a reset request, we could trigger one.

By using a DTMF (Dual Tone Multi-Frequency) decoder to translate the audible tones the phone makes when pressing buttons into the corresponding number that it represents, we could decode users' voicemail passwords when they logged in to check their voicemail over the headset.

Finally, the way these specific headsets in use at the Client worked, they broadcast at all times when powered on—not just while actively on a call. So the headset also acted as a room bug, transmitting surrounding conversations if the headset was left on, which people commonly did.

TIP

Although the Max System 8HH is no longer in production it can still be found on eBay. Additionally discone antennas is a fairly easy one to make at home. Searching online for "how to make a discone antenna" will turn up a number of tutorials explaining how to make and tune a discone antenna to a specific frequency.

The microphones could pick up conversations beyond just the wearer at times, and we could hear conversations taking place in a small office. We discovered this at another company, and used the "feature" to gain valuable information about an upcoming merger. At that company, the only people authorized to use the headsets were executives and their assistants. In essence, the company not only accidently bugged their most valuable target, they filtered out a great deal of much less interesting chatter for us.

Our Client was also interested in seeing how far away we could monitor the headsets. With the normal antenna, we could listen to the headsets from about 100 ft away, which at this site meant that we could eavesdrop from the parking lot and street around the building. Using antennas tuned for 900 MHz, (the Max Systems 8HH 900 MHz discone antenna, which sadly has been discontinued) we could easily monitor the headsets from two blocks (about 500 ft) away, showing how much performance can be increased by using a good antenna that matches the target frequency.

Remediation and Lessons Learned

As a result of the RF survey, we recommended the Client to upgrade the wireless headsets to encrypted models, or if those were too expensive, replace them with digital headsets that would be more difficult to monitor. The Client determined that the expense was not warranted given the use of the wireless feature of the headsets was not critical to operations. Most of the headsets were replaced by a wired model, which adequately remediated the vulnerability at a much lower cost. Supervisors and other workers who needed the mobility of the wireless headsets were upgraded to digital models.

To prevent inadvertent exposure in the future, the Client also implemented a process to test and approve wireless devices before they can be purchased. Using this process they generated a list of approved wireless headsets which could be purchased.

Lessons:

- *Know your equipment:* Most organizations tend to focus on the desired features that a device provides them in order to do something useful.

They don't often think about the unintended things that the equipment also does—such as broadcast a signal that can be picked up with the right equipment from hundreds of feet away further than the intended use.

- *Test your equipment.* Most of those unintended things won't be found on a feature list in a brochure. Test the security of any devices that communicate wirelessly before widespread adoption. Most organizations will think to test 802.11/WiFi devices, but don't think about things like wireless headsets, cordless phones, two-way radios, remote controls, and RFID tags.

- *Better security isn't always more expensive.* Sometimes a simple, cheap solution, such as switching to wired headsets, is better than an expensive technical solution.

- *Be aware of what your organization is saying over the air.* Avoid broadcasting critical information. If you must broadcast something sensitive on a routine basis, come up with a protocol for what to say, what not to say, and how to say what you need to (using pre-determined code words, for example) without inadvertently revealing unnecessary details.

CHAPTER 6

How to Use the Information You Gather

As organizations become aware of security weaknesses, they tend to (at least attempt to) mitigate them. What this means for the penetration tester is that assessments that would have been fairly easy a few years ago are now far more challenging. Because radio traffic is an area where most organizations do not place a focus on security, it provides a great resource for gathering intelligence. This is not just theory; the authors have used information gathered by profiling radio traffic to increase the success of many penetration assessments, as shown by the included case studies.

CONTENTS

WHO IS GUARDING THE GUARDS?

The information in this book is especially valuable when the target organization has a guard force that uses two-way radios. By listening to radio traffic on the guard frequency, it is usually possible to learn the schedule of the guard rounds as well as any regular movements. Knowing where the guards are is invaluable when you wish to avoid them. Knowing the number of guards on duty on any given shift will also give you an idea as to plan your penetration assessment. Obviously, attempting the assessment when the fewest possible number of guards is on duty is generally best. One exception is of course if there are a small number of guards during off business hours, it still may not be the best time as any movement will be seen as abnormal and will draw attention.

Learning the schedule of the guard shifts is also valuable. Note whether the shift changes are staggered, i.e. not all guards end their shifts at the same time. If all the guards change their shift at the same time, this can present an opportunity as their posts may not be as well guarded as they are in the middle of a shift. Like any of us, guards may become tired and less attentive and diligent towards the end of their shift, presenting another advantage for the penetration tester.

While monitoring traffic on the radio, write down guards' and dispatchers' names, and make note of the guards' lingo. This information is useful for social engineering. Knowing peoples' names gives the appearance that you belong, and allows you to claim that a guard directed you to go into a restricted area. Saying "the guard told me to go here" will not be as effective as saying "Bill Smith told me to go here." (Assuming, of course, that there is a guard names Bill Smith.) Peppering your conversations with guard and other organizational lingo will also give the appearance that you belong, and will help you gain the trust of people within the organization.

MONITORING PHONE CALLS

Keep in mind that legally monitoring telephone communications can be a difficult, so prior to monitoring any telephone traffic, be sure that it is legal and that you have the proper authorization. Far more sensitive information is spoken over the phone than on two-way radios, and if you are able to grab telephone traffic you will likely learn a great deal about the operations of the target organization, including names, client information, and financial data. Credit card, banking information, social security, health information, and other highly sensitive information are commonly relayed by phone. Testing whether this traffic is secure should be a priority during a physical penetration test if within scope.

For the penetration tester, being able to monitor the telephones at a help desk or call center can be the Holy Grail, especially if the help desk is involved in password resets. Gathering usernames and passwords as well as the names of people within the organization can make a penetration test fairly simple. Listening to help desk calls will give valuable insight into if and how the help

> **NOTE**
>
> Remember, criminals, by definition, do not follow the law. This means that they have tools and methods at their disposal that we, as professional security practitioners, do not. While there are many criminal methods that we cannot test without breaking the law ourselves, it is still of the utmost importance to understand the methods and techniques used by the criminals. Thinking like a criminal is often the best way to learn to defend against criminals.

desk verifies the identity of the caller, as well as organization specific technical terminology.

A DTMF decoder, described in more detail in Chapter 7 is a device that decodes the sound of telephone touch tones and displays them numerically. This is a great tool while monitoring telephone traffic as it makes it easy to grab voice-mail passwords. Additionally, some radio equipment uses DTMF codes to transmit coded messages.

WIRELESS CAMERAS

Wireless cameras are commonly found in areas where it would be difficult or expensive to run cables. Areas to look for wireless cameras include stairways, the parts of the parking lot furthest from the main building, and around the perimeter of a large property.

While the advantages of being able to capture the camera feed and view it are obvious, there is also a large amount of information that can be gathered by studying the cameras themselves. For fixed position cameras, it should be simple to determine what they are monitoring. A camera high on a light pole over a parking lot is there to provide a panoramic view, and a camera pointed towards a door is obviously monitoring that door.

Once you have gained access to the wireless camera feed, you will know what the guards can see. Note the camera coverage and look for blind spots to exploit during a physical penetration test.

Pan Tilt Zoom (PTZ) Cameras

PTZ cameras are becoming more common, and allow a user or program to physically move the camera remotely.

Follow the movement of the camera to determine if the camera is being controlled by a human or by software. If software is controlling the camera, you will see controlled and consistent movements, while if a human is controlling

> **TIP**
>
> While it is easy to zone out while watching camera feeds, there are many smart video systems on the market that detect movement or changes in the feed and alert a human, allowing them to investigate. As with all types of technology smart video systems have also gotten cheaper. Being able to alert on motion is a feature now found in many low- to mid-level video systems.

it, you will see more spontaneous movement. If the camera is being controlled by software, then it is possible that the camera feed is not being actively monitored. Conversely, if a human is controlling the camera it can almost be guaranteed that the image is being carefully watched.

View the feed at night to determine if the cameras are night vision capable, and also watch to see if there are any times of day that glare makes the camera useless. If there is a time when glare is affecting the camera, then that may be a good time to attempt physical penetration.

Note the focal length of the camera, and if it is a wide or narrow angle lens. See if the image provides usable detail. It is not uncommon that cameras are placed far enough from the target that they do not provide a great deal of information to the guard.

In general, cameras are a poor detective control. It is easy enough to lose focus while watching television, so imagine trying to pay attention while watching the feed from a bank of cameras focused on an empty parking lot. Cameras, however, are one of the best ways to determine what happened and gather evidence after a crime has occurred. As a preventative control, cameras can be useful as they may send an attacker on to a softer target.

CHAPTER 7

Basic Overview of Equipment and How it Works

In Chapter 2 the basic concept of a scanner was introduced. To review a scanner is a radio that stores multiple frequencies in channels, quickly switches between the channels, and stops scanning when a channel is active. This chapter will take a deeper dive into scanner features and operations, how to select a scanner for radio reconnaissance, and accessories that are helpful when performing radio reconnaissance.

COMMON SCANNER CONTROLS AND FEATURES

The best source of information for operating a particular scanner is the owner's manual. That being said, there are controls and features that are common to most scanners available on the new and used market.

Channels and Banks

Channels store the frequencies the scanner searches when it is scanning for radio traffic. Most scanners organize channels into banks. For example, a 200-channel scanner may have 10 banks of 20 channels each. This can be helpful when you need to group services together. A hobbyist may put police in

CONTENTS

TIP

It can save time and frustration if you program the last few banks with common frequencies. For example, the authors use bank 10 for Family Radio Service Channels, bank 9 for dot frequencies, and so on. Lower banks are then used for specific engagements and targets. This saves the need to reprogram the higher banks.

one bank, fire in a second bank, and air traffic in a third. Alternatively, if you travel, you can put channels in banks organized by city or location. It is usually simple to enable or disable an entire bank, so that instead of locking out 20 channels individually, you can simply disable a bank.

While older scanners do not allow you to change the size of the banks, many newer scanners allow you to group channels however you wish. For example, bank 1 can be channels 1–3, bank 2 can be channels 4–10, and so on. Bearcat and GRE have two competing flexible memory systems which are respectively named Object Oriented Memory and Dynamic Memory Architecture (DMA). Both systems obtain roughly the same goals and which system is better is really a matter of personal preference. If you get a scanner that uses either of these systems take the time to fully understand how the system works so you can make full use of it.

Squelch

Squelch is a control that adjusts the amount of signal needed to stop the scanner or break audio silence so that a user can listen to the transmission. If the squelch is turned all the way down, you will end up with static where there is no signal. Alternatively, turning squelch too high can cause you to miss weak signals that may be useful. The goal is to find a happy medium where you get as little static as possible while still being able to receive the desired signals. On most scanners, squelch is controlled by a knob. Some newer radios use plus and minus buttons or a rocker. While it is an issue of personal preference, the authors prefer knob control due to the granularity of adjustment it allows.

Scan Button

Pushing the scan button will cause the scanner to scan its channels for activity. Additionally, the scan button can be used to resume scanning if the channel being listened to is not of interest.

Hold Button

The hold button will stop a scanner from scanning and hold the radio on the current frequency or channel. This is helpful if you want to follow a conversation on a single frequency.

Manual Button

The manual button, also known as the direct button, allows you to directly enter a frequency you wish to listen to. This is often the first step to programming a frequency to a channel. Some scanners combine the Hold and Manual button.

Program

Program can be either a button or process followed to program a frequency to a channel. The control is different for each radio but is often similar between radios by the same manufacturer. The best source of information for instructions on using the programming function is the radio owner's manual.

Lockout Button

The lockout button is used to lock out a channel so that it is not scanned. This is especially helpful if a channel has a lot of static for some reason, or if you wish to ignore a certain channel. A common reason for using the lockout during reconnaissance is when there is another user group nearby, such as maintenance, and you are only interested in listening to other group, such as the security force. Some radios have temporary lockouts which will lock out a channel until the unit is rebooted.

Search

The search feature allows you to program in high and low frequency stops, and the scanner will scan the defined range and stop on activity. Some units will store frequencies where they find activity into scratch memory, sometimes

> **TIP**
>
> If you work for a team of penetration testers that will be sharing a scanner on or between engagements you should use a standard channel/bank layout so multiple teammates can easily use the same device. This will save time when you need to setup scanners between engagement and lessen confusion while performing an engagement.

> **TIP**
>
> When preparing to use a scanner on a new engagement always make sure all locked out channels are unlocked. This will ensure that channels locked out during the previous engagement will not be skipped at the new site.

> **TIP**
>
> If a target site has a frequency that is used to communicate with the police, it is a good idea to set that as a priority channel. During a penetration test, you want to know as soon as possible if the police have been summoned so you can be prepared to explain yourself.

referred to as "monitor channels." This feature is valuable for penetration testers, and is very helpful when looking for a signal in a set area. One option to take advantage of this feature is to set up the scanner overnight or while you are working on other things, and then return to see what the scanner has found. You can first scan the range and lockout birdies or other signals that are not of interest. This is much easier than using the search or scan button each time the scanner locks on a birdie or other undesired signal. Keep in mind that some radios will store frequencies that you locked out between power on and off cycles or searching sessions. Be sure to clear the memory before performing a search in a new location, or you may find that you have locked out the very frequency you are looking for. Some units allow selective memory deletion, meaning that you can leave some birdies in the list to be locked out. The authors, however, have found that it is easier to clear memory and start fresh.

Priority

With the priority feature, you can set a channel or group of channels as a priority. In some implementations, this channel or group of channels will then be checked more often for activity, while in others, the priority channel(s) will be checked even while the scanner is listening to another channel, and will switch to the priority channel if activity is detected. In some older implementations of the priority feature, the scanner would check the priority channel ever second, causing an annoying cutout while listening to other channels. If buying an older scanner, it is a good idea to test this feature before making your purchase to be sure it is implemented in a way you can live with.

SELECTING A SCANNER

The most important piece of equipment for radio reconnaissance is a scanner, and selecting the correct scanner can make things much easier. In fact, selecting the correct scanner can often times mean the difference between success and failure. Many folks are tempted to get a top of the line scanner with all the bells and whistles. Although this may be fun if you only occasionally use your scanner these extra features often add confusion as you need to relearn all the features each time you need to use the radio. Remember the point is to have a tool you can use and gain value from. For most targets you do not need a fancy or expensive scanner to gather valuable information. In addition to the features discussed earlier in this chapter here is some criteria to keep in mind when selecting a scanner for wireless reconnaissance work.

Form Factor

When selecting a scanner the first step is to figure out what form factor works best for your needs. There are variety of form factors scanners are available in including handheld (Figure 7.1), base scanners (Figure 7.2), and mobile scanners. Many mobile scanners can be used as base scanners, or can be mounted in a vehicle, either hardwired or powered by the car's 12 V power socket/lighter receptacle. Handheld scanners tend to be small and are battery powered, and many can use ac adapters to power the unit and charge onboard batteries. Rechargeable batteries can significantly lower operating costs if you use the scanner often. For most wireless reconnaissance work, especially the first radio, we recommend handheld scanners because they are the most portable and easy to power.

A fourth form factor that is starting to become more popular are scanners that can only be controlled using a computer. These scanners do not have any control on them besides a power switch and occasionally volume control. These radios are often just a plain metal box so are also referred to as black box radios. Because these radios require a computer to function as they are not a good first radio to get for radio reconnaissance work. These radios often contain specialty features which may be of value to penetration testers and will be discussed more in Chapter 9 New Technology and the Future of Radios in Penetration Testing.

Programmable Verse Pre-Programmed Scanners

Scanners are available both as pre-programmed units or as programmable units. The pre-programmed units are not particularly useful for the purposes of this book, and are intended for hobbyists as well as automobile racing fans and railroad enthusiasts. A pre-programmed scanner may make sense to have

---

TIP

Always be sure to bring plenty of batteries. Even if you use rechargeable batteries, be certain to have non-rechargeable batteries available in case the rechargeable batteries die. It makes for a bad day to get on site only to discover that you do not have any battery power left. Also, note that some scanners have a switch to change operation between rechargeable and non-rechargeable batteries. Be sure that the switch is set correctly.

as a second scanner that is dedicated to monitoring police traffic in the area because the radio will not need to be reprogrammed each time you perform an assessment in a new city. To be effective for reconnaissance, choose a programmable scanner which can easily be programmed in the field as new target frequencies are discovered.

Frequency Coverage

Scanners are generally divided into two basic categories as far as frequency coverage is concerned; Full Coverage and Band Coverage.

Full Coverage is generally found on higher-end units, and often includes shortwave coverage. While shortwave coverage is not particularly useful for penetration testers, it can be fun to have if you are interested in becoming a radio hobbyist. Unless you get a special radio, even a Full Coverage radio will still block old analog cellular telephone frequencies. While this may seem odd given that these analog frequencies are no longer used, it is still illegal to sell a radio that can receive these frequencies. However there are exception to this law and unblocked radios can often be legally obtained if you are willing to cut through a good deal of red tape. Contact a dealer that sells unblocked scanners and they can provide the requirements and the paperwork needed to purchase an unblocked scanner.

Band Coverage scanners cannot tune to every frequency, rather they provide access to popular bands or groups of frequencies. These are sufficient for most users, as the frequencies you will want to listen to are generally included in the Band Coverage.

It used to be that there were three basic types of radios: those banded to cover 30 MHz to roughly 500 MHz, those banded to cover 30–900 MHz and Continual Coverage. These days, more or less every scanner sold new covers 900 MHz where 900 MHz coverage used to be a differentiator, today the common differentiator is in the trunking and digital capabilities of the unit. For radio reconnaissance purposely a scanner will not need to receive signals below 30 MHz Having a scanner that can receive over 1.8 GHz is usually of limited value as well. If you are performing an assessment that requires receiving traffic over 1.8 GHz a scanner is probably not the best tool to use because most traffic in this range is digital. Chapter 11 will discuss specialized tools that are more appropriate to this task.

FIGURE 7.1 Handheld Scanner, Reprinted with Permission from Meagan Call

FIGURE 7.2 Base Scanner, Reprinted with Permission from Meagan Call

Useful Scanner Feature

In addition to the criteria provided above there are a number of other features which will be useful when using a scanner for wireless reconnaissance. These features are explained in the next section so you have an understanding of what features do and what features to look for when purchasing a scanner for wireless reconnaissance work.

Number Pad

The number pad (Figure 7.3) is used to enter channel numbers or frequencies. Not all radios will have a number pad, and require other methods to enter frequencies. Many smaller radios will forgo the number pad in the interest of saving space.

The authors prefer scanners with a full keyboard. Of course, units without keyboards are usually very small, often times the size of a standard deck of playing cards (Figure 7.4). While this can be great if you need to hide the scanner for covert listening, it can take a great deal of time should you need to enter additional frequencies, unless you are able to use a computer to program it. While these qualities make scanners without keyboards a great option for a secondary scanner, the advantages of a full keyboard scanner outweigh the small size. The full size keyboard can make or break an engagement if you forget the software, your computer, or a programming cable and need to reprogram on the fly. Programming cables can be extremely difficult to find when traveling as they are specialty items not generally carried by mainline retailers.

TIP

If purchasing an older radio, be sure to determine if it has 900 MHz coverage.

FIGURE 7.3 Scanner Number Pad

Lighted Screens and Keyboards

Lighted screens and keyboards are of course helpful in night operations and also in any low light situation you may encounter. Even if you do not plan to perform any night operations, keep in mind that even during the day, penetration testers often find themselves hiding in closets and other dark places. When selecting a scanner, keep in mind that some units have lit screens and not lit keyboards. A unit with a lit screen and keyboard is best.

Channel Spacing/Steps

Step size is the amount of adjustment when tuning to frequencies. Similar to modulation, most scanners are designed to automatically select the appropriate step size for the current band. As a general rule, for a scanner to find every signal it must tune in steps no more coarse than those used in any of the communication systems to be scanned.

TIP

In order to allow for more efficient use of the spectrum, steps in the US are changing. It is important to keep this in mind if you are looking at older units, and also important to "future proof" a newer unit. Many newer, modern units support firmware updates that can address these step changes as they occur. This does, of course, assume that the manufacturer will still be supporting the unit and releasing updates years into the future.

It is important to select a step size that matched that of the band so that you don't miss any desired transmissions. Step sizes can greatly impact search speed on a particular band. The smaller the step size, the longer the search will take, as the scanner must monitor more frequencies. The reverse is also true; larger step size will reduce search time. Most radios will automatically select the correct step size for the band, however, some higher-end radios may not select step sizes automatically, and as such you will need to memorize them or have a resource available where you can refer.

When using a search function, selecting the correct step size is critical. If you use too large of a step size, you may miss a signal and never even know it was there. Conversely, if the step size is too small you may trigger on a signal that is not at the center of the frequency, leading to a weak or distorted signal. Also keep in mind that small step sizes can dramatically increase search time. Selecting 5 kHz on the 25 kHz band will cause the scan to take five times as long. If unsure which step size to select go with the default selected by the scanner.

Scan Speed

Scanning speed refers to how quickly a unit can switch between channels while scanning. Scanning speed is generally measured in terms of how many channels per second it can scan. Ten channels and 100 channels per second are common speeds. The scanning speed ultimately affects how quickly a unit can scan a band or section of the spectrum, generally the faster the better.

Modulation

In Chapter 2 we discussed different ways radio signals can be modulated to encode data into them. Common analog modulations types supported by scanners are AM, FMN, and FMW. Single Side Band (SSB) modulation is not generally of interest to penetration testers as it is mainly used on shortwave frequencies. It is good to note, however, that if you get a high-end scanner that goes below 30 MHz, SSB support will allow you to listen to all the activity on shortwave.

Most scanners are designed to know which modulation commonly occurs on which frequency. This mean that if you tune to 146.xxx, the unit will automatically switch to FM, and if you tune to 108.xxx, the air band, it will switch to AM. Higher-end scanners will often allow the defaults to be overridden to enable

FIGURE 7.4 Icom R-6 is an Example of a Small Scanner Without a Full Keypad, Reprinted with Permission Icom America Inc. ©2012 Icom America Inc. The Icom Logo is a Registered Trademark of Icom Inc. The Use of Icom Product Images has been Approved for Tutorial Purposes

listening in non-standard situations. Being able to override the default modulation type is generally only required if the license class for the band you are trying to monitor has changed, which happens very infrequently, but having this option on a scanner is a good way to future proof it against future band plan changes.

APCO P-25 Decoding

APCO P-25 is a common digital form of modulation that is gaining popularity in public sector radio systems. Although P-25 can be used with traditional radio systems it is mainly deployed on trunked radio system. Currently P-25 is the only digital format which is decoded by consumer-grade scanners. At the time this book was published the authors have not seen any business using P-25 on their radio systems but as systems get upgraded private sector radio systems will adapt this standard. Today this is not a required feature for monitoring most business radio systems. But this will likely change in the future. Additionally this feature is useful when monitoring some police radio systems. If you can afford it the authors recommend purchasing a radio that can decode P-25 traffic so it is future proofed as this standard is rolled out in more locations.

Trunk Tracking

The concept of trunking was first introduced in Chapter 2. In general most business do not have radio systems complex enough to justify the cost of a trunked radio system. However it is not unusual for larger business to deploy trunked systems and many cities are upgrading their municipal radio system to trunked systems. For this reason we highly recommend selecting a scanner that supports trunking.

When choosing a trunk capable scanner a number of things need to be kept in mind. First be sure to find one that can trunk in the bands you will need to monitor. Keep in mind that not every trunk capable scanner can trunk in all the frequency bands it can receive. Newer trunked systems are beginning to operate in the 300 MHz and 700 MHz band, which may not be covered by older radios. New scanners that support trunking generally support trunking in all the bands the radio can receive in.

Second not all trunked systems are currently supported. It seems like every year scanner manufactures are adding support for more trunked systems to their trunking scanners. Because of this research the trunking systems supported by the scanners on the market and pick one which supports the most number of trunking systems.

Third if you purchase a trunking capable scanner be sure it is running the latest firmware. Many scanners now allow the end user to upgrade the firmware in

> **TIP**
>
> Trunked systems often have complicated programming functions. The best reference for pro-
> gramming your scanner to work with trunked systems is your scanner owner's manual and the
> RadioRefernce.com wiki.

the device to add new features or fix bugs. Frequently these bugs are related to
how the units support trunking or decoding P-25 traffic, if supported. So if you
are having trouble monitoring a trunked system with your scanner verify that
you are running the most recent firmware. If you are not, upgrade the firmware
to see if that solves the problem.

Rebanding

Rebanding, also known as reconfiguration, refers to changes in the 800 MHz
band to allow users on the Nextel Network and users on public safety networks
to coexist. If monitoring rebanded trunked systems, you must have a scanner
that supports rebanding. If you attempt to monitor rebanded trunked systems
without a scanner that supports rebanding, the scanner will go to an incorrect
channel and you will not be able to monitor the transmission. While rebanding is not usually a major concern if you are monitoring private trunked systems, it is definitely a concern when monitoring public safety systems in the
800 MHz range.

Continuous Tone-Coded Squelch System (CTCSS)

Continuous Tone-Coded Squelch System (CTCSS), also known as tone
squelch or remote squelch control, is designed to allow users on a shared frequency to hear only users in their user group. Theoretically, CTCSS will allow
a user to hear only transmissions by those in their user group, and not be subjected to the transmissions of others on the same channel. CTCSS equipped
receivers generally can operate in either CTCSS mode or normal mode. While
in CTCSS mode, the receiver's audio will only be activated if the transmitted
signal was sent using the same CTCSS tone. CTCSS is also helpful in noisy RF
environments where a scanner could pick up spurious signals that break the
squelch. By programming in a CTCSS tone for that channel the scanner will
not stop scanning unless a signal is present which also has the correct CTCSS
tone.

Digital Code Squelch (DCS)

Digital Code Squelch (DCS) is a newer digital version of CTCSS that operates
by sending digital data using subaudible tones. Some manufactures refer to
DCS and Digital Private Line (DPL).

Alpha Numeric Memory

Alpha numeric memory feature on a scanner allows the user name each channel with a text string. In the past all a scanner would show is a channel number and maybe a frequency. On those types of scanners you needed to remember that Channel 1 is the fire main dispatch for Lakewood, Channel 2 is the fire dispatch for Rocky River, and so on. When you program 200 channels into a scanner, it very quickly becomes impossible to remember them all. This feature is very helpful when performing wireless reconnaissance because you may be visiting multiple sites in a week which makes the task of memorizing channel names even harder.

Computer Programmable

It is wise to spend more money up front and acquire a scanner that can be computer programmed. Being able to program the scanner with your computer will allow you to use your full size keyboard, copy, and paste frequencies from manuals and Web sites, and save significant time if you need to enter a large number of frequencies.

Voice Squelch

Voice Squelch is a radio feature where the scanner will only stop when it has identified what seem to be voices. While this can be useful to skip over data traffic that you may encounter during reconnaissance, it will only work as well as the vendor has implemented it. It is wise to assess the scanner's ability to accurately pick out voice transmissions before using this feature on an engagement. Similarly some manufactures have a Data Skip features that skips over channels that seem to carry data signals. This feature achieves the same results as Voice Squelch.

Attenuator

Attenuation is a filter that decreases the power of signals entering into the radio. The attenuator is inline between the antenna and the front end of the radio. Various scanner models implement attenuation differently, where some allow the attenuator to be set on a per-channel basis, while others require that

---

NOTE

Recent model GRE and RadioShack scanners have a reputation of the front end of the radio easily being overloaded in environments with lots of signals, such as metropolitan areas. This results in distorted audio or hearing signals on the incorrect frequency. Enabling the Attenuation feature on the radio will help decrease the strength of these signals and prevent the radio front end from being overloaded.

the attenuator be used for all channels or none. Attenuation can be helpful in environments where there is a strong signal in the area that is overwhelming the other signals. Pager transmitters are notorious for overloading surrounding signals. Attenuation can also be helpful in an RF rich urban environment when using a sensitive scanner.

Near Field Frequency Counting Features

Chapter 6 on On Site Radio Profiling covers using a frequency counter such as the Optoelectronics Scout to find unknown frequencies used by target radio system. Some scanners have similar near field capabilities built in. Uniden has branded this technology Close Call while GRE labels it Signal Stalker. Other manufactures may have similar systems with different names. As a refresher near field receiver work by finding the strongest transmitter in the area that is high enough above the noise floor for the receiver to detect it. Generally this means these features work best in environments with low RF noise where you are close to the target transmitter. The usability and usefulness of these systems can vary greatly between models, manufactures, and the operating environment. Overall the authors have had mixed results with using these built in features and much prefers to use a dedicated device such as the Optoelectronic Scout. However if your scanner does have these features it is worth learning how to use them and employing it during assessments. There have been rare cases were the near field frequency counter in a scanner detects a signal the Scout could not detect. The more information you can gather while profiling a target the better.

Discriminator Out

When a scanner receives a signal it goes through a number of filters and processes to clean up the audio before you hear it. When listening to normal audio traffic this is a good thing so you get clean and clear sound which is easy to understand. However during this cleanup process information is lost that is needed to decode digital signals. Some high-end scanners have a discriminator out feature that outputs this data before it has been cleaned up so digital signals can be decoded. Decoding these digital signals is beyond the scope of this book, but this feature may be of interest to individuals who perform wireless reconnaissance if they plan or want to try to intercept and decode signal signals. This is also a feature that can be added to most scanners by the end user. However usually this involves opening the radio and soldering connectors to the internal circuit boards. So needless to say this is not for the faint of heart and will void your warranty. If you attempt this the normal warning apply and the authors of this book will not be held responsible for any damage you do to your radio. Figure 7.5 shows the inside of a radio with a homemade discriminator out added.

FIGURE 7.5 Scanner with a DIY Discriminator Out Added, Reprinted with Permission from Meagan Call

Additional Considerations When Buying Used or Older Model Scanners

There are many arguments to be made for purchasing used or older model equipment, price being chief amongst them. Used units are also great for

NOTE

Ham radio stores are usually a great place to purchase used equipment. Often times, they will offer some form of warranty, and if something goes wrong you will be able to contact the person who sold the equipment. Additionally, Hams are passionate about radio, and generous with their knowledge.

TIP

If funds allow, it is best to have two scanners. One can be used to monitor dispatch and the other can be used to scan. On reconnaissance engagements, you will almost always find at least two things you will want to monitor.

physical penetration tests; it is preferable to drop a $100 used unit in a stream instead of a brand new $500 unit. There are of course additional consideration when buying used equipment. Used equipment is often buyer beware, and assessing the remaining service life is beyond the scope of this book. Do, however, buy from a reputable seller and learn about the unit so you can judge its merits and condition.

SCANNERS RECOMMENDED FOR WIRELESS RECONNAISSANCE

Even with laying out the critical used to select a good scanner for wireless reconnaissance some individuals may prefer specific recommendations on which radio to buy. The next section contains these recommendation on currently produced scanners that would work well for radio reconnaissance. Note that new scanners come out yearly so very quickly this list could be out of date. Also because a radio is not listed in this section does not mean it's a poor fit.

Uniden Bearcat BCD-396XT

Currently the top scanner we recommend for wireless reconnaissance is the Uniden Bearcat BCD-396XT (Figure 7.6). This is the scanner most frequently used by the authors and during most of the case studies in this book. This scanner covers the frequency range that encompasses all the commonly encountered during wireless reconnaissance. The scanner can follow trunked systems and decode P25 digital audio traffic so it is as future proof as you can get with current production scanners. The 396XT comes with Uniden's Close Call feature which can be helpful when trying to identify unknown frequencies. The scanner is also computer programmable and controllable. Currently the BCD-395XT retails for about $450.

GRE PSR-310

If you don't want to decode P25 digital traffic or if the price of the BCD-395XT is outside your budget the next model down to look at is the GRE PSR-310 (Figure 7.7). The PSR-310 frequency coverage is almost identical to the BCD-395XT and will also track analog trunked systems. The PSR-310 includes GRE's Spectrum Sweeper feature which can be used to find the frequency of unknown close transmitters. The scanner is also computer

FIGURE 7.6 Uniden Bearcat BCD-396XT, Reprinted with Permission from Meagan Call

FIGURE 7.7 GRE PSR-310, Reprinted with Permission from Spencer E Holtaway

programmable and controllable. At time of publication the PRS-310 is selling for around $150.

AOR 8200MKIII

The AOR 8200 MKIII (Figure 7.8) is a high-end continual coverage scanner that has a number of unique features which can be helpful to individuals performing wireless reconnaissance. The big shortcomings of the AOR 8200 MKIII are it lacks the ability to track trunked radio systems and cannot decode P25. Although it cannot decode-25 digital signals it can demodulate all popular analog signals. A unique accessory this scanner has is the ability to decode voice inverted traffic using the AOR VI8200 Voice Inverter Card. Using this card the radio can decode single point voice inversion transmissions. The radio is also very easy to control using a computer. AOR has published the full specifications on how to control the radio so someone with programming ability can write simple programs to control the scanner. The case study at the end of Chapter X used an computer controlled AOR 8200 MKIII to take signal strength readings from multiple locations to determine which wireless headsets were inside the target building. The computer control port on this scanner also allows easy to access the discriminator output for the radio which is helpful if you need to decode or analyze data transmission on frequency. The 8200 MKIII can also be paired with an Optoelectronic Scout to perform Reaction Tuning where the radio automatically tunes into the frequency that Scout detects. This can be a valuable feature to quickly determine if the frequency the Scout discovered belongs to the target organization. With all these features also comes one of the biggest disadvantages of the 8200 MKIII. It is a very complicated radio that has a steep learning curve. If you do get one for wireless reconnaissance work be sure to acquaint yourself with the radio and its features before you take it out in the field. Also be sure to bring along the owners manual in case you need to look

FIGURE 7.8 AOR 8200MKlll, Reprinted with Permission from AOR USA Inc.

something up. At the time of publication the AOR 8200 MKIII sells for around $750. In general the AOR 8200 MKIII makes a good second or third radio for individuals performing wireless reconnaissance. Another good second or third radio is the Icom IC-R3 which is a handheld scanner that can decode video signals. Although the IC-R3 is no longer produced it can still be found used on eBay and at ham radio stores.

BUILDING YOU KIT: HELPFUL ACCESSORIES

Before heading out, the authors will make sure that they have the right equipment for the job. If you are planning to use radio reconnaissance in a penetration test (and hopefully you are, if you are reading this book) it is helpful to have the following equipment. The first and most important accessory is getting a quality antenna.

Antenna Connectors

Most scanners use either BNC or SMA antenna connectors. When selecting an antenna for your scanner make sure you get one that has the appropriate

> **TIP**
>
> If we had to choose between a cheap radio and a good antenna or a good radio and cheap antenna, the authors of this book would always choose a good antenna with a cheap radio. A good antenna will pull in signals better so that a cheap radio can hear it. However a cheap antenna will make a good radio deaf.

connector or get the appropriate adaptor. Also do not confuse the SMA connector used on scanners with the reverse-SMA connection used on some 802.11 wireless cards. An SMA and a reverse-SMA connector are not compatible. Figures 7.9 and 7.10 show what these various connectors look like.

Antennas for Handheld Scanners

Every new handheld scanner will come with an antenna. These free antennas usually work OK to poor and should be one of the first upgrades when building a wireless reconnaissance tool kit.

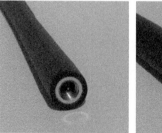

FIGURE 7.9 Male SMA, BNC, and Reverse-SMA Antenna Connectors, Reprinted with Permission from Meagan Call

FIGURE 7.10 Female SMA, BNC, and Reverse-SMA Antenna Connectors, Reprinted with Permission from Meagan Call

Flexible "Rubber Duck" Antennas

The flexible "rubber duck" antenna (Figure 7.11) offers many benefits; foremost among these benefits is convenience. These antennas are fairly easy to conceal, are rugged, and as you may have guessed, are flexible. Rubber ducks are also fairly inexpensive, making them an ideal choice for environments where your radio equipment will be subject to abuse. It is far better to break a cheap antenna than an expensive one.

Rubber duck antennas also have their disadvantages. As a general rule, they have poorer range than other varieties of antennas. Be especially aware of small

FIGURE 7.11 Various Rubber Duck Antennas, Reprinted with Permission from Meagan Call

> **NOTE**
>
> Once bent a collapsible whip antenna is difficult if not impossible to bend them back into their original shape. For this reason, it is best to buy a new one.

1-in. rubber duck antennas. While compact, and neat looking, they generally perform very poorly unless used in high frequencies in the 900 MHz range and above.

Telescoping Whip Antennas

Telescoping whip antennas (Figure 7.12) can be tuned to the desired frequency by shortening or lengthening them. They are generally very good for receiving signals; however they do have several disadvantages. They are not particularly rugged, and are fairly easy to bend or break during use. Their larger size makes them generally more difficult to conceal, which is at cross purpose for our intended uses. They are also rigid, making them a poor choice when wearing a radio on your belt. If not careful, rigid antennas can also place strain on the radio and cause the internal antenna connector to break from the circuit board. If you plan to use a telescoping whip with a portable radio, choose one with a joint so that it can be pointed skyward when the radio is laying flat. This will be a great help if using the radio at a table, desk, or in your car.

Recommended Handheld Scanner Antennas

If you can only get one antenna, we recommend a high-quality rubber duck antenna such as the Austin Condor or Diamond RH77CA. Both are good multi-band antennas. Beware of rubber duck antennas that are very short, around an inch in size. Although these small antennas may be easy to conceal, oftentimes they have very poor performance and will cause you to miss everything except very strong signals. If you need a small antenna you can easily conceal try the Diamond RHF10 which is only 2.75 in. high, very flexible, and receives 450 MHz, 800 MHz, and 900 MHz signals surprisingly well for its small size.

If possible we highly recommend getting at least two scanner antennas. The second antenna should be a telescoping whip. The algorithm discussed in the Antenna Theory section of Chapter 2 can be used to figure out how long or short a telescoping whip antenna should be to match the target frequency. If you need to listen to a specific frequency, investing in a frequency-specific antenna may be a good option.

FIGURE 7.12 Telescoping Whip Antenna, Reprinted with Permission from Meagan Call

> ### TIP
>
> Antennas are sensitive to all metallic objects in the nearby area. Moving an antenna often has a dramatic effect. This is especially true in high metal cubicle farms, inside office buildings, hotel rooms, and the like. If using radio equipment while in a vehicle, simply moving the car a few inches may be enough to improve the reception. Of course some times, you will need to find another location altogether to get a clear signal.

Mobile Antennas

For mobile antennas, the authors generally recommend the magnetic mount type. This allows easy portability and can be used on multiple vehicles. If you don't need to use the antenna on more than one vehicle, a permanent mount antenna is of course also an option. In a pinch, the RadioShack mobile antenna is a good option, and it is usually fairly easy to locate a RadioShack. Better options include the Larson Tri-Band (Figure 7.13) and the Austin Spectra antennas. The Austin Spectra is the taller and more expensive of the two. Both of these antennas are quality units, and both have their hardcore devotees in the radio community. The Larson and the Austin antennas can both be used with an NMO magnetic mount (Figure 7.15). NMO stands for New Motorola and is a type of antenna connector commonly used on mobile antennas. A BNC magnetic mount (Figure 7.14) can be used to mount a portable antenna on the outside of a vehicle, however this may not be the best option for long-term durability because BNC connector are not water tight.

Coax Cable

Coax cables are a special type of cable used to carry RF signals. The quality and type of coax cable used become more important the longer the cable needs to be. For most wireless reconnaissance work long coax cables are not needed. Therefore most of the details around coax cables and selecting a high-quality cable are purposely not included in this book. For the purposes of wireless reconnaissance the most important thing to remember is to match the impendence of the cable with the impendence of the antenna and receive. Scanners and scanner antennas have an impedance of 50 ohms.

DTMF Decoder

DTMF decoders decode telephone touch tones. If cordless telephones or headsets are used by the target, DTMF decoders can be useful for decoding voicemail passwords. Some high-end scanners have DTMF decoders build into them. DTMF decoders are also available as an external piece of hardware or software.

FIGURE 7.13 Larson Tri-Band Antenna, Reprinted with Permission from Meagan Call

Camera

As discussed in Chapter 5 Onsite Profiling having a camera can be helpful for taking pictures of the radio equipment and antennas used to by the target organization so they can be analyzed later. A camera with a good optical zoom will be of the most benefit. Additionally binoculars can be helpful for similar reasons.

FIGURE 7.14 Magnetic Antenna Mount for Antennas with a BNC Connector, Reprinted with Permission from Meagan Call

FIGURE 7.15 Magnetic Antenna Mount for Antennas with a NMO Connector, Reprinted with Permission from Meagan Call

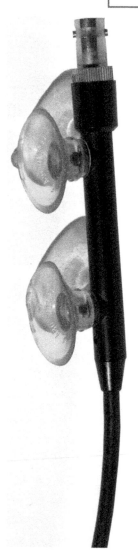

Headphones or External Speakers

Headphones are another must have accessory when performing wireless reconnaissance. They will allow you to discretely list to radio traffic. The crackle of a police radio dispatcher will always gather attention in public. When working in a car an external speaker can help hear transmission over road and engine noise. When getting a speaker to use in a car also consider getting one with a build in audio amplifier so you can turn up the volume loud enough to clearly hear radio traffic.

Audio Recording Equipment

Being able to record information you are receiving is helpful in case a transmission is missed, if a transmission needs to be reviewed to gather additional information or if you want to record activity on a channel while you are away from the radio. Also recording and playing back the information leaked by a company can be a great way to illustrate the risks to upper management of the intercepted data. Many desktop and mobile scanners have a line out designed specifically for being hooked up to a recording device. If you have a handheld scanner you can also connect the recorder right into the headphone jack. When selecting an audio recorder pick on that has a build in VOX feature that will start recording only when audio is present. Finally, although not critical, try to select a recorder that will time stamp the audio files so you can look up with the transmission occurred. Note that some scanners such as the Uniden Home Patrol, GRE PSR-800 and Icom R-20 (Figure 7.17) come with built in sound recording capabilities.

Video Decoder

The ability to decode video signals given off by a target organization is immensely helpful. Many organizations will use wireless security cameras to monitor area where they cannot easily run a

FIGURE 7.16 Suction Cup Windows Antenna Mount, Reprinted with Permission from Meagan Call

video cable. Some high-end scanners such as the Icom IC-R9500 (Figure 7.18) have the ability decode video signals. However radios like this are outside of most budgets. Other options include getting a used Icom IC-R3 which is a handheld scanner that can decode video signals. Although no longer in production used models can be readily found on eBay and other online sources. A dedicated video interception device such as the Optoelectronics Video Sweeper or AOR AR-STV (Figure 7.19) is another option. These are self-contained devices explicitly designed to scan for and decode video signals. Note that most of the units will only decode analog video signals. In Chapter 10 we'll touch on how the upcoming technology of software defined radios can also be used to decode both analog and digital video signals.

RF Amplifiers

In general the authors of this book do not recommend using RF amplifiers (Figure 7.20), also known as antenna boosters or pre-amplifiers. An RF amplifier will amplify not only the signal you are trying to hear but also other RF noise in the area. Generally use of an RF amplifier will just overload the front end of the scanner and cause you to miss the signal you are searching for.

Voice Inversion Decoder

Warning: Be sure to check applicable state, local, and federal laws before attempting to decode transmissions using voice inversion.

Voice inversion is a simple way to scramble audio transmissions to make them difficult to understand. Voice inversion technology can be broken down into two types single point inversion and dual point

FIGURE 7.17 Icom R-20 with Build-In Sound Recording Capabilities, Reprinted with Permission Icom America Inc. ©2012 Icom America Inc. The Icom Logo is a Registered Trademark of Icom Inc. The Use of Icom Product Images has been Approved for Tutorial Purposes

FIGURE 7.18 Icom R-9500 Has a Build In Video Decoder, Reprinted with Permission Icom America Inc. ©2012 Icom America Inc. The Icom Logo is a Registered Trademark of Icom Inc. The Use of Icom Product Images has been Approved for Tutorial Purposes

FIGURE 7.19 AOR AR-STV, Reprinted with Permission from AOR USA Inc.

inversion. Single point inversion flips the audio at a set point. Dual point flips the audio at a set point and flips it again at a different point. Breaking a single point voice inversion system is fairly simple because only one value needs to be brute forced. Breaking dual point inversion is more difficult, but still not impossible, because both inversion points need to be discovered.

FIGURE 7.20 RF Amplifier, Reprinted with Permission from Meagan Call

It's important to note that neither dual point nor single point voice conversion are actually encryption. They are simply ways to scramble the signal. Voice inverted transmissions are pretty easy to spot because they sound like Donald Duck speaking gibberish.

Some older scanner had voice inversion decoders built into them such as the Alinco DJ-X2 and DJ-X2000E. These units could only decode single point voice inversion traffic and are generally very difficult to find on the used market. As discussed earlier the AOR 8200 MKIII is capable of decoding single point voice inversion with the optional VI8200 Voice Inverter Card installed. Ramsey Electronics makes the SS70C Speech Scrambler/Descrambler Kits which can decode some single point voice inversion traffic. If you search the web you will also find a number of programs which claim to decode voice inversion traffic. Overall the authors of this book have found these programs to be unreliable and high suspect, so use them with caution.

In general very few organizations still use voice inversion technology to protect their transmissions. If the organization has a strong need to protect their radio traffic they have moved to either digital systems with a custom vocoder or implemented a true encryption technology. However the ability to decode voice inverted traffic is still helpful if you encounter a target using this feature.

CHAPTER 8

The House Doesn't Always Win: A Wireless Reconnaissance Case Study

INTRODUCTION

After all the technical information, and all the studying, the authors sincerely hope that the reader will use the information in this book in their career as a professional penetration tester. We have used the techniques and information presented in this book on multiple penetration assessments, and often these techniques allowed us to be successful where we may not have been without them. The following story is, well, mostly true. Because we are professional security people, we cannot disclose details of our clients' systems nor their names. All the events in this case study happened, however we drew from events from multiple penetration assessments to create this scenario.

OFFICE WORK

It would be nice to say that our story began on a cold dark night fraught with danger, clichéd as that may be, but it didn't. Our story begins in a nondescript, modern office where the temperature was a comfortable climate controlled 71 °F. We had been contacted by a casino a few states over; the administrators had read a few articles about hacking, and they were concerned as to their exposure. Casinos have a long history of top notch physical security, and they know almost all the tricks. But even though many of the games look the same as they have for decades, the whirring gears and mechanical innards have been replaced by chips and wires. The muscular pit boss standing cross armed behind a dealer was long ago supplemented by CCTV, the "eye in the sky," however those cameras are now transmitting through the air rather than across an easily guarded cable.

CONTENTS

We went through the standard things, contracts and such, ensuring that we had all the legal aspects in order, and then had our kick-off meeting with our client. We were given nearly free reign over casino and its systems. While those are often the most enjoyable assessments, allowing us to use all the tools in our arsenal, we were also very cognizant of the fact that unlike many of our physical penetration assessments where getting caught meant facing fellow geeks and explaining ourselves to the police, the casino was guarded by big men with guns. Preparation was essential to ensuring not only that we would be able to uncover and identify weaknesses in the casino's security, allowing them to then mitigate these weaknesses, preparation was also key to ensuring our own physical well being.

Although we were itching to get to the exciting part (who doesn't enjoy dressing up in dark clothing and legally being able to attempt to break into a casino?) there were hours of preparation ahead before we could even consider attempting our breach.

Settled in front of our computers with plenty of coffee, we began our preparation using the sources that have been described in this book. Google maps gave us the satellite images and street view images that allowed us to map points of entry and even determine where external guardhouses were located. Digging into radio enthusiast Web sites, however, turned up gold.

On RadioReference.com we searched for both the casino name as well as the city in which the casino is located. Searching for the city gave us the channels for both the local police as well as the county sheriffs' department. Venues such as casinos often have sworn law enforcement assigned onsite full time, often

---

TIP

Remember there are other online mapping sources beyond Google. These other services sometime contain different images or features which may be useful. For example Bing Maps offers a Bird's Eye view which gives you a three-dimensional view of the building. This feature can be useful for figuring out how tall a building is, number of entrances and where the entrances are located.

TIP

If you get lucky and find that RadioReference.com or another site has listed the target frequency, it can save you some effort. Another method, when looking for frequencies in a far away location, is to post on radio enthusiast message boards what you are looking for. Keep in mind, however, that this may tip your hand.

Sheriffs deputies. Searching for the casino name, however, we came up empty handed.

We scoured the Internet looking for press releases and other information that could tell us what types and brands of radios were in use by the casino guard force. In this search, we were not as lucky. This just meant that we would need to perform a little more onsite reconnaissance.

OUT IN THE FIELD

After packing up our equipment, and making sure that we had our "get out of jail letter" in hand, we hit the road in earnest.

The drive was easy, clear roads, and a clear night. A few hours later, we were in the parking lot of the casino. We immediately began to monitor the law enforcement frequencies that we found on RadioReference.com, and discovered that they were correct. Remember, even though you may have discovered the frequency prior to arriving onsite, you must always verify that it is correct. Sometimes people make mistakes, and sometimes frequencies change. We sat in the parking lot for a few hours listening to the scanner. It quickly became apparent that the police dispatcher and the casino guard dispatcher enjoyed talking about everything, work related or not, over the open frequency. We learned the names of several second and third shift guards, as well as many other random facts about them. One guard had a daughter that recently graduated from high school, and was planning a party. Another guard was notorious for leaving his post unattended. Some of the information we gathered would prove useful and relevant, while other information would prove to just be noise. We kept notes of everything we heard. Even though we knew we would remember a lot of the information, we still wrote everything down. The casino was undergoing a renovation, nothing too major, new carpet and some new paint. As the night wore on, we noticed contractors driving up to the gate of the employee lot in pickups and vans, and being waved in. Clearly, they were working in the overnight hours when the casino was less crowded to make sure they disturbed as few patrons as possible.

> **NOTE**
>
> A "get out of jail letter" is a common term in the security industry for the letter the target company will give you to describe the scope of the penetration test. If, during the course of the assessment, you find yourself mistaken for a real criminal, you can give the letter to law enforcement to prove that you are not a burglar. Of course, don't be surprised if you end up in handcuffs while the police contact the target company and verify your information and mission.

NOTE

It is a good idea to periodically remind personnel who use radios that everything they say on the radio can be heard by anyone with a scanner. In fact, you often don't need a scanner to listen to emergency dispatch frequencies. Smartphone applications such as 5-0 Radio and similar allow users to listen to scanner channels worldwide. While monitoring public channels we have often heard discussions that could prove embarrassing for those using radios, and of course, we hear a lot of information leaking. Of course radio traffic can be encrypted to protect the privacy of the communications, however in reality very few agencies or businesses encrypt their radio traffic.

As it got later, and the crowds in the casino began to disperse, we decided to call it a day and retire to our hotel. Without the crowds, we began to stand out in the parking lot, and didn't want to get caught before we even started. Of course, had we been concerned about being detected while in the casino parking lot, we could have found a less conspicuous place to park, such as one of the businesses or hotels nearby. In the hotel we compared notes and discussed our plan for the next day. Rife with anticipation, we bedded down for the night.

GLITZ AND GLAMOUR

The night before, we took note of how the casino patrons tended to dress. Most seemed casual, so we were sure to mirror this. Had we been in an upscale casino, we would likely have dressed like high rollers. Remember, it is about fitting in and not seeming out of place. We parked the car, walked through the lot, and through the front door. We took note of the locations of the guard stations as well as emergency exits and doors marked "employees only." We settled into seats in front of a few penny slot machines, and placed our backpack containing a Scout on the floor next to us and started pulling the lever. While we didn't win big on the slots, we hit the jackpot when a guard walked past and we noticed that he is carrying a Motorola HT Pro Series radio, most likely a GP-338. These radios operate in 29.7–42, 136–174, and 403–470 MHz, and do not support encryption or trunking. With this information, we smiled knowing that we could monitor these radios if we could find their frequency. The radios used by the guards had a 4–6 in. antennas, which suggested that they operated in the 136–174 or possibly the 403–407 MHz range. We left the slot machines and wandered through the casino, our Scout in the backpack doing its work. We spent a good hour in the casino buffet, getting every penny's worth from our $14.95 all you can eat special. Just another perk of the job.

With full stomachs, we returned to the car and began to pull the frequencies from the Scout. We focused on the 136–174 and 403–407 MHz range we believed they operated in. We programmed this small group of frequencies

NOTE

While not necessarily as portrayed in Hollywood, casinos don't mess around. They are constantly on the lookout for cheaters and thieves, and any time you are in a casino, you can pretty safely assume that you are on closed circuit television. The night before, we noticed that many casino goers carried backpacks, and thus knew that we could carry one without drawing too much suspicion. Had the job required it, we could have used a fanny pack or other method to conceal our Scout. Just know that if you are caught in a casino with an electronic device, you will have some explaining to do. Like any other penetration assessment, know what will allow you to fit in and what will arouse suspicion.

into our scanner and began to monitor. It wasn't long before we found the frequencies used by the casino guards.

LEARNING THE LOCAL LINGO

Over the next few hours we learned a great deal about the casino and how it was protected. In addition to general chatter (the coffee is cold, my team lost the softball game), we heard the guards beginning and ending their shifts. We learned the names and nicknames of the guards, their schedules and post locations, as well as all sorts of lingo used by the casino. It was early afternoon, and we had what we would need, as far as radio reconnaissance goes. We headed back to our hotel to rest and regroup.

TIME TO GAMBLE

We put our plan together, deciding that we would impersonate construction contractors and attempt to enter the employee entrance. We knew we had to look the part, and brand new work clothes wouldn't work. We found a local thrift shop and purchased some tired looking workpants and shirts marked with the stains of construction. After a light dinner and a quick nap, we left the hotel and drove to the casino parking lot just after 2 AM, the time we noticed that the contractors were showing up for work. We sat in the car, sipping coffee, waiting for our opening. It wasn't long before we heard the guard at the employee lot gate pleading with the dispatcher to leave post so he could use the restroom. As the guard's requests became more desperate, and there was no replacement available at the time, the dispatcher finally relented. She told him to be quick, and that no replacement was necessary as he would "only be a minute." A minute was all we needed. We quickly drove through the unattended gate and into the employee lot. We knew that the gate was covered by a surveillance camera, however we decided to take a chance. We had a trunk full

> **NOTE**
>
> Social Engineering is one of the most effective tools in the arsenal of the penetration tester. Whether social engineering on the phone to get a network password, dropping infected USB sticks in a lobby, or working your way past a guard, trust is key. Radio reconnaissance, as we have shown in this case study, is a great way to gather information that can be used to gain trust.

of equipment. Wire cutters for the fence, carpet to throw over the barbed wire if we had to scale it, but it would remain in the trunk, all because the guard had to use the restroom, and we were listening to our scanner. We parked the car, grabbed our bags, and walked right in the employee entrance. We informed the guard at the desk by the employee entrance that we were there to complete a job. We dropped the name of the dispatcher, asking how she was, and how her daughter Riley was doing. All information we had learned by monitoring the guard channel in the scanner. The guard smiled, and handed us our visitor badges after we signed the visitor log. As we clipped our visitor badges to our shirts, we made the transformation from trespasser to authorized visitors, with the credentials to prove it.

We enjoy picking locks and scaling fences as much as anyone in the security industry, but it is much easier to have an authorized person open the gate and unlock the door for you. We didn't have to pick locks or break any doors, we just had to take advantage of the knowledge we gathered during reconnaissance. People are helpful, no one wants to give a hard working contractor a hard time. Give them a reason to trust you, and they will. We walked right in.

INSIDE

While inside the casino, we made sure to use an earpiece and monitor the guard and local police channels so we would know if we had been detected. It is always better to know when the police are coming, so you can be prepared to give them your get out of jail free card, and ensure that your hands are clearly visible. We wouldn't need that this night, however. What follows isn't so much about radio reconnaissance as it is about penetration testing, but it demonstrates the rich reward that awaited us.

We found a conference room and sat down. We installed a wireless access point, hidden in the ceiling tiles, so we could continue to attack the network from outside. We unrolled a length of CAT-5 and plugged our laptop into a wall jack. We knew what we were looking for. We targeted the gaming network, security systems, credit card processing systems, and the data warehouse containing customer information. This casino was ahead of the curve. They knew all about convergence, all their physical security devices were on the corporate network. Instead of sneaking around, ducking cameras, and hiding in shadows, we just needed our laptop on their network.

We wandered around the limited access areas of the casino, saying hello to other contractors and guards as we passed them. We tried our visitor badges on various doors, and found that they gave us access to various janitor closets and other low value targets. We didn't attempt to go near the counting room or any other target that would likely have drawn suspicion and blown our cover. Back in the conference room, we gained access to the video system. The video system used the default manufacturer password, which we knew. Manufacturer default passwords are available on the manufacturer Web sites. We learned which areas were covered by surveillance and which ones weren't. We even watched the guards watching their screens. We knew exactly where the managers were, where the pit bosses were, and where the guards were.

While pilfering through a file share, we found a spreadsheet showing which slot machines and electronic games were hot. We could have used this information about machines with high payout rates to play certain games and increase our odds. We found a ton of customer information too. We had the names, addresses, descriptions, and notes on all the high rollers. The publicity from this type of break in could cause monetary damage to the casino far worse than cleaning out the counting room. We found information on other customers too—thousands of them. Information that would be invaluable to the marketing efforts of competing casinos.

A little later, we heard the door to the conference room open, and saw a large angry looking man standing in the doorway. We knew we had been caught. The man wasn't angry that we were there though. He was angry that we were sitting in the conference room when there was work to be done. He sternly told us to get back to work, and that he would let us slide this one time. As he escorted us out of the conference room, and pointed to an area that needed to be painted, we knew our work was done. We left through the door we had entered, said good night to the guard, and went to the hotel to sleep.

CHAPTER 9

New Technology

As our world gets smarter, it seems to also get more wireless. This means greater exposure to risk and more opportunities for the radio frequency security researcher. Looking to the future of RF security, two things will define the future of radio hacking and wireless reconnaissance:

- Everything is going digital.
- Software-defined radios (SDRs).

EVERYTHING IS GOING DIGITAL

Currently, there is still a lot of analog radio transmission, and there likely will be for some time to come, but increasingly radios are moving to digital. Currently, some scanners support decoding P25 but that's really the extent of the digital formats they can decode. It is likely that in time, more scanners will support limited TDMA formats. TDMA is a general digital format, used by some older cellphones, and is also starting to be deployed by some radio systems. Motorola has a standard built on TDMA called "P25 Motorola X2-TDMA" which is the upgrade to the current P25 systems. Motorola's X2-TDMA system was released before Phase 2 requirements were finalized and is largely based on Phase 2 standard. So far, one scanner supports this standard: the GRE PRS-800—currently support for X2-TDMA is in the testing stage. MOTOTRBO is another TDMA-based digital radio standard made by Motorola. Currently no scanner can decode this.

The other standards are up in the air. Many of them are closed standards and the manufactures do not want to license the ability to monitor them to scanner makers. Many use the fact that scanners cannot monitor them in their marketing as a security feature. This does work to keep the general public from being able to casually listen in—for now. Of course, projects like DSD and

CONTENTS

GNU Radio are working to reverse-engineer these systems and create tools to monitor them. Digital Speech Decoder (DSD) is an open source software package that works with GNU Radio, which can decode some of these formats. For more on this, see the section on GNU Radio later in this chapter. There are many digital formats to be aware of, however, among them: Terrestrial Trunked Radio (TETRA), Vector Sum Excited Linear Prediction (VSELP), OpenSky, iDen, EDACS ProVoice, and DECT, as well as numerous proprietary standards.

TETRA is popular in Europe (in fact it is sometimes known as Trans-European Trunked Radio), but if it's used in the United States, we haven't seen it. VSELP is an older digital standard for digital encoding of speech, which is close to being phased out.

The undeniable trend in electronics over the past 20–30 years is that more and more stuff is going wireless. These days, it is more than just the IEEE 802.11 (WiFi) protocols, which is what most penetration testers are currently focusing on. There are lots of other standards—and proprietary wireless formats as well. As these grow in marketshare and mature, it will be increasingly important to be aware of their presence or potential presence in the environment.

Beyond 802.11—Digital Wireless Protocols

Digital Electronic Cordless Telecommunications (DECT)

DECT is a format mainly used by cordless phones and headsets, where it is becoming the *de facto* standard, but may be found in use for other applications as well, such as baby monitors. Toolsets like DEDECT were built to intercept and attack these systems. The current implementation of the DEDECT requires a Dosch Amand COM-ON-AIR Type 2 PCMCIA Card (Figure 9.1), which is

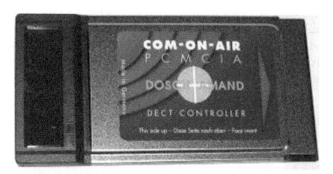

FIGURE 9.1 Dosch Amand COM-ON-AIR Card

currently very difficult to find. Information on the DEDECT toolset can be found at https://dedected.org/.

DECT supports encryption in the form of a proprietary algorithm called the DECT Standard Cipher (DSC). Through research by the security community, it has been discovered that the encryption was poorly implemented, due to poor random number generation and inadequate key length, and the security algorithm has been broken.

In many cases, the encryption can be disabled on DECT devices by injecting a spoofed data packet that appears to be from the other unit, saying that it does not support encryption. Failing back to cleartext transmission, it now broadcasts unencrypted, leaving the communication exposed. This can even occur mid-call.

DECT's broken-by-design security model is a good example of a vulnerability that went unnoticed until people started investigating them. What problems have gone unnoticed in the protocols and systems below because security researchers have not looked at them and security was not thought about from the start?

IMPORTANCE OF ASSESSING WIRELESS PRODUCTS BEFORE THEY ARE IMPLEMENTED

This is why it is important to research new radio technology before implementing it. Without fail, proprietary encryption schemes are weaker than publically tested and reviewed encryption, because the public systems can be peer-reviewed by top encryption experts, and flaws that are exposed publicly can be corrected, or at least avoided.

Security assessments are key when implementing a radio system. Do not trust that encryption is strong, or even enabled. If you're engineering a new technology, involve security researchers and testers in the design process from the beginning. If you're shopping for a system, do your homework *before* you buy. Start out by conducting a risk assessment to determine what your risks are, and determine what level of security is needed. If at all possible, perform a vulnerability assessment on products and systems you are considering buying *before* committing to them, especially if your application for the system is sensitive or mission critical. Different organizations will have different security needs. If you do need encryption, avoid proprietary algorithms, and demand a trusted, open standard such as Triple DES or AES. Having the ability to apply firmware updates is a great feature that more equipments should utilize to patch problems

when they occur. We are seeing more devices starting to support this. Customers should keep in mind how easy or difficult the firmware updates are to apply, however. Many devices support firmware updates, but may need to be sent back to the vendor or manufacturer in the worst cases. Even in the best cases, often you must visit each device and manually apply the update, which is very time consuming.

The advanced two-way radio systems and smartphones do allow for patching over the air, which simplifies the matter, but this opens up a potential new vulnerability: rogue firmware injection over the air. Rogue firmware injection is currently more of a concern with devices where physical access is needed to install firmware, but it is a theoretical exploit vector that should be assessed. It's important to look at how the device verifies the firmware that is loaded onto it, to prevent someone from loading malicious firmware. Again, a thorough risk assessment will determine whether this is a threat to the specific organization. Creating and loading a custom firmware is currently a sophisticated attack that many companies probably don't need to worry about, given other vulnerabilities in their environment.

Bluetooth

Bluetooth is an extremely common technology, found in pretty much every cellphone, most laptops, many desktop-class personal computers, and in ever-growing number of cars. Bluetooth uses frequency-hopping spread spectrum in the 2.4 GHz ISM band, the same band used by WiFi, microwave ovens, and most other 2.4 GHz consumer devices. Bluetooth devices form a piconet containing up to eight nodes—one master and seven slaves. Bluetooth is used for short-range communications between device peers, as well as device to peripheral. The number and variety of peripherals which communicate via Bluetooth is immense—wireless headsets for hands-free cell phone use, keyboards, mice, videogame controllers, audio speakers, you name it. Bluetooth is best known for transmitting audio, as in Bluetooth headsets. But it can also be used to connect HID devices such as keyboards and mice, and send data. Although not very popular, there are Bluetooth access points which function the same as WiFi access points to connect multiple devices to a network.

Attack tools against Bluetooth exist, such as Ubertooth (Figure 9.2), but still Bluetooth is not commonly targeted in penetration tests. Ubertooth is a custom made radio dongle that can attack radio systems in the 2.4 GHz range. Originally it was created to attack only Bluetooth, hence the name, but has since been expanded. The project's home page is http://ubertooth.source-forge.net/.

Ubertooth can be used to monitor traffic, inject traffic, and do basic spectrum monitoring. It is very much a platform still in development, so new features are being added all the time. It is one of the cheaper, if not cheapest, ways to sniff Bluetooth, and the cheapest tool to inject custom packets. It has lowered the cost of entry to start attacking Bluetooth devices.

The key with Ubertooth is it is very difficult to take a consumer Bluetooth dongle and have it sniff and inject custom frames. With WiFi, this was very easy to do. Nearly any WiFi adaptor can be used to sniff traffic, and today most support injection as well. Once this was discovered, it became a lot cheaper to attack WiFi. To date, the only way we have seen to accomplish this on a Bluetooth dongle is to load a commercial firmware, which has probable EULA

FIGURE 9.2 Ubertooth One Dongle. Reprinted with Permission from Meagan Call

violation implications. Commercial tools to sniff and inject Bluetooth packets cost thousands of dollars. Ubertooth has lowered the cost for a device to attack Bluetooth to $120.

Zigbee

Zigbee is a specification based on IEEE 802.15.4 standard for Personal Area Networks, and typically used in short-range and low-power wireless devices. Commonly, Zigbee devices are used in mesh networks, to extend their range by relaying through intermediary devices in the mesh. It is used in a wide range of products from smart meters to thermostats to wireless light switches. Zigbee shows up in a surprising number of places when you look for it, but very few penetration testers are looking for it. Zigbee radio systems are engineered into a wide variety of applications.

The KillerBee framework was developed by Josh Wright to attack Zigbee. The project's home page is: http://code.google.com/p/killerbee/. It is free, open source software. KillerBee is a framework for assessing and exploiting vulnerabilities in ZigBee and IEEE 802.15.4 networks. It can be used to sniff packets, replay packets for reply attacks, and to attack the crypto system used by Zigbee. Figure 9.3 shows KillerBee being used to inject custom crafted packets using Scapy.

KillerBee can be used with a couple of different Zigbee radios. The most popular is the Atmel RZUSBSTICK (Figure 9.4). KillerBee has limited functionality when using this radio with the stock firmware. By loading a custom firmware more features are available. Note that loading the custom firmware does require

HIGH-JACKING WIRELESS CONNECTIONS TO GAIN ACCESS TO INTERNAL SYSTEMS

In all cases, the wireless links are potential attack points, and should be considered by penetration testers. By attacking the links or devices, most times the attacker is looking at manipulating the systems they control. For example, you may be able to make it look like more power is being used. Or attack the end point, like a thermostat, and adjust the temperature in the building. This may not seem like a big deal, but for manufacturing and healthcare, temperature control is critical.

However, if you did penetrate a device you may be able to use that to further penetrate into the backend corporate network. For example, on a recent penetration test of a smart grid system we were able to high-jack the radio link between the smart meter and the power company. The power company assumed this was a secure link, and left the rest of their network much more exposed than they should have. So once we were on it, we had full access to their backend systems, and could execute a more traditional penetration test and attack the systems the smart meters talked to. We did find vulnerabilities that allowed us to take over these systems and use them as a jumping-off point into the larger network.

FIGURE 9.3 Using KillerBee to Inject Custom Crafted Packets. Printed with Permission from Spencer "zeroSteiner" McIntyre

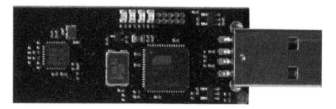

FIGURE 9.4 Atmel RZUSBSTICK Used by KillerBee. Reprinted with Permission from Meagan Call

an ATMEL programming tool ($200 USD). But overall, it is still cheap to create a custom attack tool. The big benefit of KillerBee is it created a cheap platform to sniff and inject packets to attack Zigbee from. In the past, test gear to do this was very expensive. With KillerBee, you can take a $40 Atmel RZUSBSTICK and make it into an attack tool.

Digital Means More than Just WiFi

Clearly, there is a great deal more in the world of digital radio than just 802.11 (WiFi). Some example applications for non-WiFi wireless technologies are:

- *Home/building automation systems:* Common brands are Crestron, Control4, and X10. They are used to connect together lighting, HVAC, motion and temperature sensors, motorized window shades, and other appliances, so buildings and homes can be automated and more energy efficient. For example, the automation system can be programmed to adjust the HVAC system to turn on and off, depending on which rooms are occupied, and the weather forecast for the day. All of those light switches, sensors, thermostats, etc. are often connected wirelessly. Some use Zigbee or custom implementations of 802.14.4, while others use proprietary RF technology. These systems are not only green, they can also save money, so they will become more popular, particularly as prices on the technology drop and energy costs rise.

- *Smart Grid:* The most popular smart grid applications currently are with electric power grids, but can apply to gas and water distribution networks as well. The technology is used to create a grid that can adapt in real time to changes. In a basic implementation, smart meters only report usage data back to a central office in real time, or near real time, but lacks the ability to control. To be truly smart, the grid has to be able to take the information it has gathered, and change/react to better manage load.

  Smart grid meters can interact with home appliances to adjust energy usage. For example, they can interact with thermostats to set a higher AC temperature during the hot season if the grid is under heavy load. Or, they can talk to smart appliances, such as laundry machines and dishwashers, to schedule them to run when energy costs are the cheapest, or when there is a surplus, and avoid running when energy is in high demand. The meters, appliances, and the infrastructure to support this are often wirelessly connected using a variety of protocols, from proprietary mesh networks, to Zigbee, to cellular.

- *Proprietary links for digital signs, billboards, etc:* Generally, billboards that have full color displays are updated over cellular links so they get faster speeds. Lower-tech signs like the gas price signs or the LED signs outside of banks and such use proprietary radio links that are usually 9600 Baud. They are slow, but not much data needs to be sent. There are really no standards here. A sign made by company A usually can't be controlled by a controller made by company B.

SOFTWARE-DEFINED RADIOS (SDRs)

We are accustomed to think of radios as hardware equipment, made of transistors, circuitry, and an antenna. However, it is possible in theory to virtualize any type of electronics hardware by emulating it in software. Thus, the software-defined radio (SDR) is born. The general design of an SDR is an RF front end which feeds into an analog-to-digital converter (ADC). This digital signal is fed into a general purpose computer for processing (Figure 9.5). In some

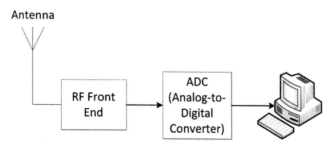

Antenna

RF Front End

ADC (Analog-to-Digital Converter)

FIGURE 9.5 Typical Software-Defined Radio Block Diagram

designs a digital signal processor (DSP) is placed between the ADC and general purpose computer to aid in processing the output from the ADC. The RF front end varies, but converts the RF signal to a level and format usable by the Analog-to-Digital converter. In a perfect world, you could just plug an antenna into an ADC, but in reality you need some front-end hardware to convert the RF to a useable format and level.

SDR is not a new concept, but with FPGAs and other electronics advances, it is now cheaper to implement them. Also, thanks to the speed of desktop computers, it has become feasible to implement features in software that formerly required specialized hardware for performance reasons. With the rise of cheaper SDRs, we are entering a new golden age of wireless hacking. While many SDRs are still expensive, prices are continually dropping, as with all electronics. SDRs will be the key to intercepting, decoding, and hacking the new digital RF systems.

How Does an SDR Differ from a Traditional Radio?

Software emulation of features traditionally implemented in hardware. Components that are usually implemented in hardware in a traditional radio are implemented in software in an SDR. The software and processing can run on a general purpose computer (personal computer or embedded system). Some use Field Programmable Gate Arrays (FPGAs) to assist with processing. Often a combination of both FPGA and a general-purpose CPU is used.

Common parts of the radio implemented in software:

- *Filters:* Making hardware filters that adapt to changes in the signal is very complicated and requires a large number of components. This is much simpler in software because the filter can be changed easily, based off of characteristics of the signal. Because of this, filters are commonly implemented in software. All audio filters in an SDR are implemented in software. Many times RF filters are also implemented in SDRs.

- *Amplifiers:* Due to the laws of physics, hardware amplifiers always introduce noise when they amplify a signal. Better hardware amplifiers have less noise, but noise will still be introduced. Software can amplify a signal without introducing noise. The bigger issue here is making sure the signal was sampled at a high enough rate so it can be amplified without losing resolution/clarity—generally speaking, the higher the sample rate, the better the result. It should be noted that in most SDRs today, the RF amplifier in the front end is implemented in hardware, not in software. However on higher-end radios, more of the front end is implemented in software, and as SDRs advance these benefits will be realized by more readers.
- *Modulators/demodulators:* One of the most common parts implemented in software. The other components in this list are often still implemented in hardware for many models of SDR. It is much easier to add new demodulators in software. It is also easier for software to give the end user the capability to adjust the modulators manually, or automatically, depending on the signal. Building a hardware demodulator that has the same level of flexibility and control is fairly complicated, and therefore an expensive proposition.

Advantages of SDR

- *Cost:* By emulating costly specialized hardware components, manufacturers are able to make radios with fewer, and cheaper non-specialized components.
- *Rapid development:* The great advantage of SDR is you can make changes to the radio by changing code instead of re-wiring a hardware component. This allows for fast prototyping. If problems exist, you seldom need to re-wire the hardware; you just need to patch the software. Once one developer produces and distributes the code, anyone can download it and run it. With a physical radio there is skill needed to make the radio which can decode the signals, once someone figures out how to do that.
- *Upgradeability:* With most obsolete hardware, usually all you can do is replace it. A lot of radio hardware still works perfectly fine, but lacks features or refinements that new models offer. With obsolete software, you can easily upgrade it, and gain new features and functionality.
- *Flexibility:* SDR provides far more features and configuration options than you can fit into a single piece of radio hardware.
- *Device consolidation:* One piece of well-made and versatile SDR hardware can be adapted to do multiple things by loading in new code. In the past, it was necessary to build multiple, purpose-built radios—which got large and expensive. Today, you can modify a single SDR to perform multiple

functions based on the code it is running. The flexibility of SDR means we can gain the ability to decode video and other data formats such as POCSAG (pagers) by writing new programs. For example in open source SDR, such as GNU Radio, there is existing source code for decoding two-level and four-level FSK signals. This functionality often costs extra in commercial SDR products.

- *Signal capture and advanced analysis.* Finally, one of the greatest advantages of SDR is the ease with which RF signals can be recorded and captured into a file. This is great because you can perform more extensive analysis on a recorded signal than you can with a live signal. Traditional radios require a signal to be present in order to perform analysis on it, but with a recorded signal you can go back repeatedly and analyze it in different ways. This is a huge help when analyzing an unknown signal. With traditional radios, at best you could record the audio output of the radio and analyze it later.

THE POWER OF RECORDING RAW RF SIGNALS

Offline Analysis of Unknown Data Signals

For much the same reason professional photographers like to shoot in RAW format, it's always worthwhile to capture the original, unprocessed RF signal whenever you can. For analyzing signals, recording is incredibly useful, because you no longer need to analyze the system live, which can involve lots of time waiting for a signal. Once you capture the transmission, you can take it off site and analyze as much as you like. Being able to replay the sample signal multiple times is hugely beneficial.

On one project, we needed to reverse-engineer the radio signal used to control a PLC (Programmable Logic Controller) which is used to control an industrial process. Once we found out the frequency of the signal, we could hear data when we tuned into that channel and decoded it as an FM signal. When we fed this signal into a visualization tool we could see it looked like a square wave. We might have been able to try to decode the binary signal right there. However, at the time we were not sure how the control signal was encoded over the air, whether it was FM encoded, or it was just lucky that it looked like an FM signal. So we captured the raw RF signal, and the decoded FM audio signal.

When we later analyzed the RF signal, we found out it was an FSK encoded signal, and could only be decoded from the RF capture. If we had attempted to capture the signal as just FM, we would have wasted time, and been forced to return to re-capture the signal, if we could get another opportunity.

The raw capture can also be helpful when analyzing frequency-hopping systems. When frequency hopping is occurring it can be very difficult to guess the hopping pattern and sync up with the signal in real time. However if you capture the raw RF signal, it is easier to analyze the hopping pattern off-line. Additionally, if you do figure out the hopping pattern you can always feed the raw RF capture back into GNU Radio and decode the traffic. Note when doing this you need to make sure you capture all the signal. So if the hopping system you are analyzing hops over a 5 MHz range, for example, you want to make sure the raw capture is at least 5 MHz wide.

WHAT IS AN FSK?

FSK stands for Frequency-Shift Keying and is a common digital modulation used to transmit information wirelessly. To decode FSK signals you need an FSK decoder. Decoders come in two types: 2-level and 4-level. These decoders are also called data slicers.

A 2-level decoder is a fairly simple circuit to build in hardware (Figure 9.6), and can decode traffic up to 2400 Baud, although 1200 Baud is a more realistic upper limit. A 4-level decoder is a more complicated circuit to build, but still possible for the hobbyist, and can decode faster data rates, and will also decode 2-level signals.

A hardware FSK decoder takes the audio output from a radio and converts it to serial (i.e. RS-232, or RS-232 over USB), which can be fed into a computer so a program on the computer can process it. The decoder does not need to know what type of data it is decoding.

FSK decoders, with the appropriate software, can be used to decode POCGAG and Flex, which are two popular protocols used by pagers, Trunking Control channels, and APCO-25 traffic.

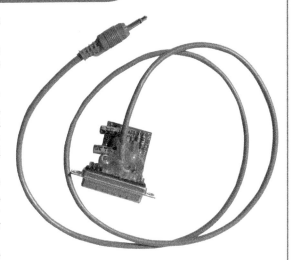

FIGURE 9.6 Hardware 2-Level FSK Decoder. Reprinted with Permission from Meagan Call

Unless you know exactly how to decode the signal, we recommend capturing the RF signal so you can process it later on. If you know how to decode the RF signal to a usable audio format you can do that as well. If the RF capture is done right (capturing the entire signal) you have the raw signal with no processing. Once the original RF signal has been processed into an audio format, if the audio signal was processed wrong (AM versus FM for example) the data needed to decode the signal could be lost.

Usually an SDR will interface with the host PC via a USB cable, but some high performance units use high speed connections such as gigabit Ethernet.

Disadvantages of SDR

- Depending on what you want to do, it may require both software programming abilities and a deep understanding of RF and electronics. This can even be a concern when customizing someone else's code.
- Data decoding features are either expensive (commercial products), or else rough around the edges (open source software still in early development).
- Some SDR hardware can be very expensive.

GNU Radio

GNU Radio is an open source toolkit for building software-defined radios. It is released under the GPLv3 license, and features powerful signal processing software, signal processing blocks built in C++, connected together by glue code written in Python, called "graphs." It contains a number of pre-built blocks and graphics to decode a wide range of things, from broadcast FM radio to GSM cell phone traffic. Figure 9.7 is a graphical representation of a Wide Band FM radio written for GNU Radio.

GNU Radio is fairly hardware independent. The Ettus Research Universal Software Radio Peripheral (USRP) is probably the most popular radio that can be used with it. Cost for a complete USRP starts around $800–1000 for a USRP1 and a daughter card. The price range is mainly determined by which model of daughter card you get.

To set up Ettus USRP, there is a main chassis that supports different daughter boards, which allow the radio to transmit and receive on different frequencies. There are different USRP models available, designed to allow processing of different amounts of bandwidth, some offering onboard processing power, and available in USB or Ethernet interfaces. The USRP1 (Figure 9.8) is currently the most popular model with hobbyist and penetration testers. It uses a USB 2.0 interface, and can support up to four daughter cards. It supports up to 8 MS/s to the host PC. Depending on the daughter board, it can receive and transmit over a very wide frequency range. The USRP1 currently costs $700 without any daughter cards.

Currently Ettus makes 13 different daughter cards for the USRP. Four of the daughter cards are popular with penetration testers, so will be discussed

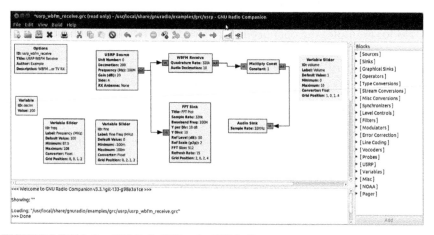

FIGURE 9.7 Wide Band FM Radio Written for GNU Radio

FIGURE 9.8 Ettus USPR1. Reprinted with Permission from Meagan Call

here. The WBX daughter card (Figure 9.9) can transmit and receive between 50 and 2200 MHz and costs $400. The TVRX2 is a receive-only daughter card that covers 50–860 MHz and costs $200. The TVRX2 card can also receive two signals at once. Due to the frequency ranges they cover, the WBX and TVRX2

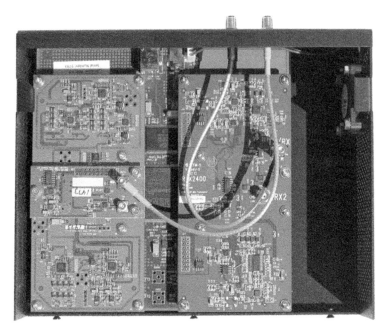

FIGURE 9.9 Inside of an Ettus USRP1 with the WBX and RFX2400 Daughter Cards Installed. Reprinted with Permission from Meagan Call

cards can receive most signals of interest to penetration testers performing wireless reconnaissance.

The RFX2400 and XCVR2450 are two daughter cards which are also popular with penetration testers. The RFX2400 (Figure 9.9) can transmit and receive in the 2.3–2.9 GHz frequency range which covers the frequencies used by 802.11b/g, Zigbee, and Bluetooth. The RFX2400 costs $275. The XCVR2450 expands on the capabilities of the RFX2400 by adding the ability to transmit and receive in the 4.9–5.9 GHz band. This allows the card to operate in the frequency range used by 802.11a, 802.11n in greenfield mode, and cordless phones operating in the 5 GHz band. The XCVR24500 costs $400. If a penetration tester has the budget, the authors recommend investing in a USRP1 chasse with the WBX and XCVR2450 daughter cards.

Example: Using the GNU Radio as a Spectrum Analyzer

Spectrum analyzers are helpful tools to visualize a chunk of the RF spectrum. This can be helpful when trying to find new signals or seeing how a signal behaves. For example, this function can be helpful to spot frequency-hopping spread spectrum radios. Issuing the command *usrp_fft.py -f* followed by the frequency will open a new window which is seen in Figure 9.10.

Using GNU Radio a $20 USB TV Tuner

GNU Radio also supports a variety of inexpensive USB TV tuners. These usually cost around $20. Look for compatible ones containing the Realtex RTL2832 chipset. Though not nearly as powerful as the USPR, they cannot transmit, are limited in their ability to process bandwidth, and receiver performance can be poor, but for $20 it's a great toy to experience the power of SDR, and its weaknesses may be lessened as better software is developed for the radio.

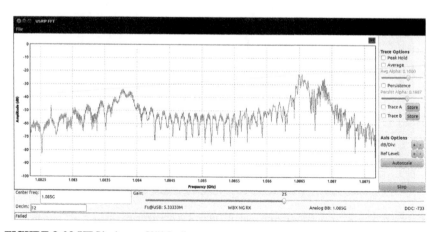

FIGURE 9.10 FFT Display on GNU Radio

Dongles with the Elonics E400 tuner have the best performance discovered to date. They can tune in to frequencies from 64 to 1700 MHz, with some gaps around 1100–1250 MHz (since that frequency band was not in use for television broadcast). They have a theoretical sample rate of 3.2 MS/s, but as of this writing, they only deliver up to 2.8 MS/s. Third-party drivers are needed to get them to work with GNU Radio. Gz-baz and osmocom rtl sdr are two projects that build these drivers.

WiNRADiO

WiNRADiO is one of the most popular consumer SDRs. It is a commercially developed group of SDRs, and consists of receivers only. They come pre-built, so there is no assembly required, and start at around $700 and up. Figure 9.11 is the WiNRADiO G305e receiver. The WiNRADiO software that comes with these radios provides the ability to decode all analog formats (AM, FM, SSB). Adding capability to decode digital formats requires additional software, which costs extra. WiNRADiO has a polished GUI interface (Figure 9.12), especially when compared to the interface on GNU Radio. Most of the WiNRADiO software is written for Windows, but there is some support for Mac OS X and Linux.

WiNRADiO offers a Professional Demodulator option for $200 which makes it easy to see how signals are demodulated for the various formats it

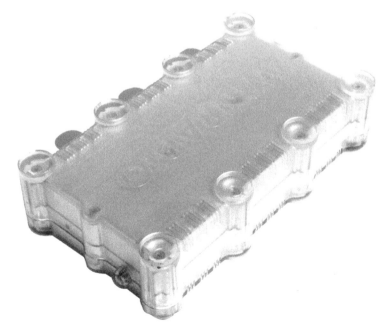

FIGURE 9.11 WiNRADiO G305e Receiver. Reprinted with Permission from Meagan Call

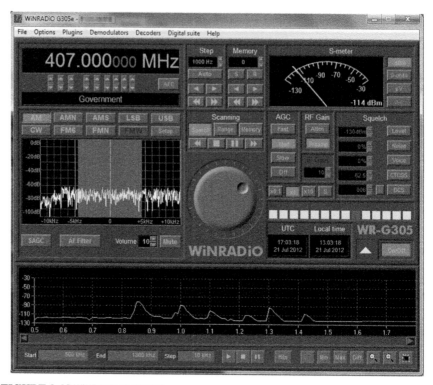

FIGURE 9.12 WiNRADiO G305 GUI

supports, and to adjust the filters and processing applied during the demodulation process. When the Demodulator is in Professional Mode, set under the Demodulator menu on the main menu bar, you can open a tool to analyze and customize how the signal is demodulated. To open this tool click Setup in the demodulator section of the WiNRADiO GUI. On this initial screen (Figure 9.13) a number of variables can be adjusted, which affect how the signal is demodulated.

The audio filter settings can easily be adjusted by clicking and dragging the edge limits of the filter. Figure 9.14 shows the default filter setting and Figure 9.15 shows the filter after it has been adjusted.

Additionally, on this screen the demodulation settings can be adjusted to determine how the signal is demodulated. To assist with adjusting the demodulation settings, a block diagram of the demodulation structure can be displayed (Figure 9.16) by clicking on the *View Demodulation Structure* button on the *Demodulator Settings* window. By clicking on the various points in the block diagram you can see what the signal looks like as it is processed by the

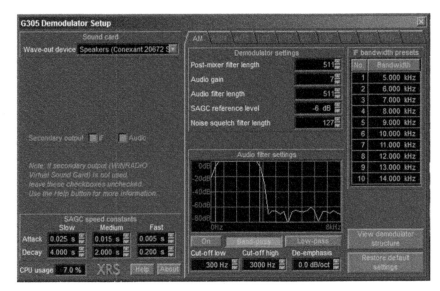

FIGURE 9.13 WiNRADiO Professional Demodulator Setup Window

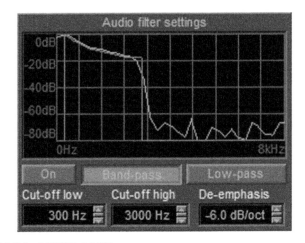

FIGURE 9.14 Default FMN Notch Filter

demodulator. This view is helpful when fine-tuning a demodulator, to see how various settings affect the processing of a signal.

The capabilities of a WiNRADiO can be extended via a plug-in architecture called XRS (eXtensible Radio Specification), which allows individuals to write their own plug-ins, or use plug-ins created by a third party. Most of these plug-ins are free, but some cost money. They allow users to add functionality to fit

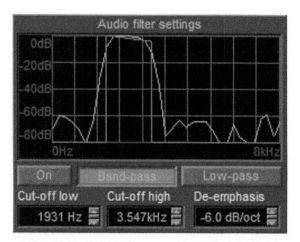

FIGURE 9.15 Customized FMN Notch Filter

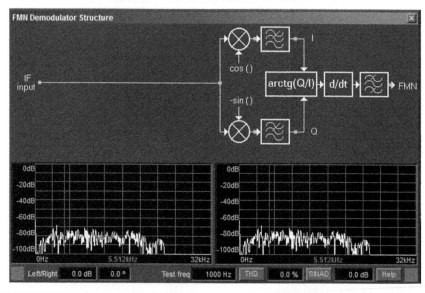

FIGURE 9.16 Demodulator Block Diagram

their needs. There is still not a large community writing WiNRADiO plug-in—far more people are creating new blocks and graphs for GNU Radio. WiNRADiO sells a number of plug-ins to expand their products. There is also an online site at http://xrs.WiNRADiO.com where third-party developers can post their XRS extensions. Some popular examples of these extensions include:

- *APCO P25 Decoders* ($99.95–$199.95 depending on the radio), made by WiNRADiO. Note this just decodes the APCO P25 digital format to voice, and does provide any trunk the tracking capabilities. (Trunking is provided by a separate software package; see below.)

- *Trunking software* ($99.95–$199.95, depending on the radio it is created for), made by WiNRADiO. This package allows you to track trunking radio systems as they switch channels. (See the section on Trunking in Chapter 2 for more information on how trunked radio systems work.)

- *Advanced Digital Suite* ($199.95), made by WiNRADiO. It decodes WEFAX, HF Fax, NAVTEX, Packet Radio, ACARS, CTCSS, and DTMF. CTCSS and DTMF are the most useful for radio recon. It also has some signal analysis capabilities, which can be useful for identifying the encoding used on an unknown data signal. One of the more challenging parts of signal analysis when you find an unknown data signal is figuring out how it is encoded. By listening to it, you can tell its data—it will sound like a modem, and with experience, it's possible to make a guess at what it is by how it sounds. For example, pagers sound different from trunking control channels. Bearing identification by ear, the signal analysis tools in the Digital Suite can do the job. For example, using an audio water fall graph (Figure 9.17), it's possible to see patterns in the signals which could be a clue to its format. An oscilloscope view (Figure 9.18) can help you figure out if it is FSK, simple on/off keying, or another data format. FSK signals have a distinct looking waveform. So if you see that pattern, the next step is to feed the signal into an FSK decoder, see what comes out, and try to figure out how to decode that.

- *Universal FSK Decoder* ($499.95), made by WiNRADiO. This can decode a number of formats but the most useful one is the ability to output raw

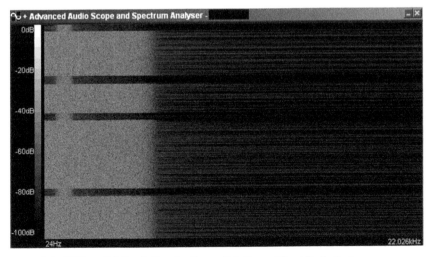

FIGURE 9.17 Water Fall Graph Showing Transmit Pattern of Burst Radio System

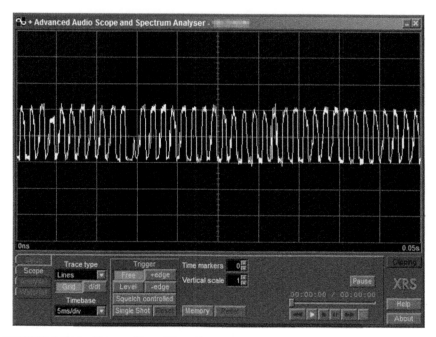

FIGURE 9.18 Oscilloscope View of Data Signal

bits which can be processed by external programs. Analysis tools are very helpful when decoding signals, especially non-standard formats. Its FSK is only a two-level decoder, so formats that require a four-level FSK cannot be decoded. Many modern FSK formats use four-level encoding. So this is a pretty big weakness especially for a $500 piece of software.

- *Mini SIGINT* (Figure 9.19) is a free plug-in that takes a finger print of a section of spectrum, and then alerts when changes over a certain threshold appears in that spectrum. This tool is useful for finding new signals. For example, if you know a device works in a certain band, but do not know the exact frequency it is using, you could take a baseline scan of that band to obtain a fingerprint of the ambient RF, then power up the unknown device and re-scan. The plug-in will identify what frequencies appeared when the device was active.
- *Band Search* (Figure 9.20) is a free plug-in that allows you to easily select a band to search for activity.
- *Waterfall Scanner* (Figure 9.21) is a free XRS plug-in which graphs the activity of a set frequency range in three-dimensions (signal strength versus frequency versus time). The plug-in can also log this data to a txt or CVS file, which is helpful for off-line analysis. Note this plug-in does not log raw RF data like GNU Radio can. Instead it logs the data used to create the graph such as the frequency, signal strength and time of each signal

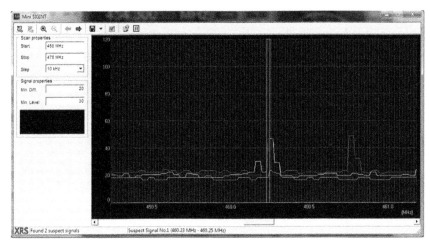

FIGURE 9.19 Mini SIGNIT XRS Plug-In Identifying New Signals

Band Search		

Band Description	Frequency [MHz]	XRS
Private land mobile service	896.000	901.000
Private land mobile service	896.000	901.000
Reserved	901.000	902.000
Amateur band (33-centimetre)	902.000	928.000
Private fixed services	928.000	929.000
Paging	929.000	932.000
Fixed services	932.000	935.000
Private land mobile service	935.000	940.000
Reserved	940.000	941.000
Fixed services	941.000	944.000
Auxiliary broadcasts and satellite links	944.000	952.000

FIGURE 9.20 XRS Band Search Plug-In

was detected. This tool is helpful for visualizing the activity of a band over time, and determine if seemingly random signals are following a pattern.

WiNRADiO is excellent for finding new signals and performing analysis on digital signals before decoding. However, unless the right decoder exists for WiNRADiO, and we have purchased it, we usually decode them with GNU Radio in the USRP. The polished interface of WiNRADiO makes it better for

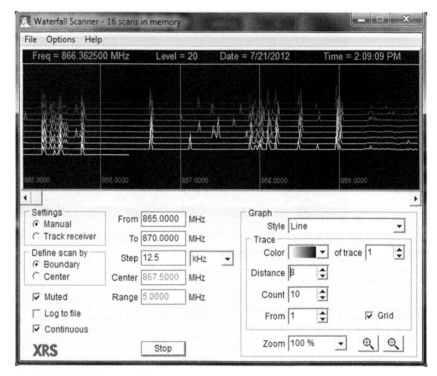

FIGURE 9.21 XRS Waterfall Scanner

quickly detecting a signal, and from there we can determine if it is of interest and perform further analysis using GNU Radio, which in many cases can do things that cannot be done with WiNRADiO, thanks to the support of the open source community.

There are other consumer grade software-defined radios, but most cover HF bands which are of limited use in wireless recon so are not discussed here.

NETWORK-ENABLED DISPATCH SYSTEMS

Network-enabled dispatch systems are a recent newcomer to the environment, and are beginning to draw the interest of penetration testers. In the old days, the dispatcher needed to be close to the transmitter to dispatch units. This limited where a dispatcher could be physically located. Everything was hardwired, which also limited the mobility of the dispatcher. The hardwired nature of these systems also offered some security.

High-end, modern data radio dispatch systems often operate over IP-enabled networks using VOIP. One benefit is that dispatchers can now move around to any place they can get an IP connection. The downside is the systems are now on IP-enabled networks, which open them up to new attacks (or old attacks, depending on your perspective).

Case Study: VOIP-Enabled Dispatch Radio on an Open TCP/IP Network

A University recently updated to a new IP-enabled dispatch system. Initially, they were very excited because this allowed the dispatcher to easily move around the campus. With a laptop and some other portable equipment, she could plug-in anywhere on campus and start dispatching. They foresaw that this would afford their security great disaster recovery capabilities if the security building was taken over or they could not operate from there for some reason.

Not understanding the potential vulnerability they would be exposing themselves to, they opened up access to the radio system so the dispatcher could be on the teacher, student, or any of the campus wireless networks. Unfortunately, this meant that anyone with access to these fairly public networks could attack their new radio system. In looking at the radio system, we found the software needed to interact with it was difficult to find, and we could not find it online. This offered them some protection, as long as this remained true, at any rate. The system wasn't designed to be put on a public network, which the student network basically was, and was poorly secured because of that. All of the traffic generated by the dispatch system was transmitted in cleartext, including user authentication, VOIP streams, and information about unit assignment. The dispatch system had an activity log, where the dispatcher could note which units are responding to what calls, and when they finish with the calls. This sensitive information would have been highly useful to unfriendlies, who could have easily used the public networks at the university to monitor the dispatcher's coordination efforts, and use the information to keep one step ahead of the security force, or spoof the system to misdirect units to where they weren't needed.

Additionally, the hardware what was provided and supported by the vendor was not well secured. For example, the system was missing security patches. The database running on the system had a very weak SA password set. The system had a large number of listening services which exposed additional attack surface. The large number of open ports wouldn't be alarming on an internal network, but were far too many for a system connected to a public network anyone could join.

First, we recommended placing this system into its own private DMZ. The vendor would not apply patches until the system was thoroughly tested. So there would be long periods of time where the system could not be patched.

We recommended setting up a VLAN on the campus teacher network that could access this VLAN and assign a few network ports spread across campus that had access to this special VLAN and therefore the radio dispatch system. These special network jacks were also placed in secure locations where not anyone could plug-in and use them. So in an emergency they know they could go to these locations to gain access to the dispatch system. This VLAN was also given a high Quality of Service (QoS) priority on the network, to ensure the dispatch traffic was not dropped when the network got congested.

We debated using a VPN but the vendor would not guarantee it would work over a VPN, given the latency it could introduce and how sensitive VOIP can be to latency issues.

The vendor also said they do not recommend accessing the system over a wireless link because of the potential latency issues. So that, paired with the security concerns, stopped them from using it on the wireless.

CONCLUSIONS AND LOOKING FORWARD

In summary, there's a great deal more to wireless than just 802.11. There is a clear trend toward everything changing over to digital formats, and moving away from analog. Wireless communications are becoming increasingly commonplace just about everywhere—not just to replace existing wired systems, but adding "smart" capabilities to devices and systems which previously lacked them. There is a great deal of opportunity here for the penetration tester, as new technologies are implemented without due consideration given to the security implications. An ever-increasing surface area for potential vulnerabilities to be discovered will keep security professionals busy for years to come.

Software-defined radios are bringing about a golden age of wireless hacking. Their ability to reduce the amount of specialized hardware that you need to carry with you, combined with its superior flexibility and adaptability, and the enhanced signal analysis capabilities that it brings to the table give the security professional a very flexible and powerful tool.

With radio, as with any high tech field, the only constant is change. It's important to keep your knowledge of the industry current. Staying abreast of the state of the art is a challenge. The best way to do that is to become immersed in the community. Read trade and hobbyist magazines such as *Monitoring Times* and *Popular Communications* on a regular basis. Getting your amateur radio license and joining local radio and scanner clubs are both great ways to get started with building your hands-on experience. American Radio Relay League's (ARRL) Website (http://www.arrl.org/) is a great source of information on how to get your amateur radio license or locate a local amateur radio club. The radio community is also very active on the Internet. Radioreference.com provides community forums, which are a source of both news and lively discussion, and provides opportunities to make contacts and friends with other professionals in the RF field.

Glossary

A

Active Antenna: Describes a high gain antenna that is short in length. These are useful where space is at a premium and commonly used on shortwave frequencies below 30 MHz.

ADC: Analog to Digital Converter.

AGC: *Automatic gain control.*

AM, Amplitude Modulation: Varying, by modulation, the transmitter power output in line with the modulating audio signal variation.

Amplification: The process by which the strength of a signal is increased. Both radio and audio signals can be amplified.

Attenuator: Circuit that reduces receiver sensitivity in fixed steps. These fixed steps are measured in decibels.

Attenuation: Blocking radio waves.

Attenuation Coefficient: The amount of power absorbed by a dielectric. A low attenuation coefficient will allow radio waves to easily pass through the material.

B

Band Pass Filter: Filter that allows a specified ranges of frequencies through, while rejecting those that are not within the specified range. Usually constructed by combining a high pass and low pass filter.

Bandwidth: The frequency space which is occupied by a radio signal.

Bank: A scanner bank is a way to organize channels. They allow for grouping a set of channels into logical banks.

Baud: Measured in bits per second, the rate data is transmitted.

Beam Antenna: Antenna, used outdoors, that receives best from a certain direction.

Beat Frequency Oscillator, BFO: Circuit in a receiver that enables reception of SSB signals by generating a replacement carrier.

BCB: *See Broadcast Band.*

Birdie: Birdies are spurious radio signals created by the internal electronics in a scanner. These signals have no sound or static however will stop the scanner in its search mode. The average scanner will have four or five birdies.

BNC: Common antenna connector which locks in place by being twisted.

Broadcast Band, BCB: Also known as AM, BCB is the frequency range between 540 and 1700 kHz. This is the AM band on your car radio.

C

Carrier: A radio transmitter's unmodulated output.

Center Frequency: In an FM transmitter, the unmodulated carrier frequency.

Channel: Frequency where radio transmission occurs. May also describe the input and output frequencies of a repeater or the location where a frequency is stored in a scanner's memory.

Coaxial cable, Coax: An electrical cable specifically designed to transmit RF energy. In a coax cable the inner conductor surrounded by a flexible insulating layer which is then surrounded by a tubular conducting shield. Invented in 1880.

Cochannel Interference: When frequencies adjacent to the signal cause undesired interference.

Communications Act of 1934: The Communications Act of 1934 combined and organized federal regulation of telephone, telegraph, and radio communications. The Act created the Federal Communications Commission (FCC) to oversee and regulate these industries. As new communications technologies have been created, such as broadcast, cable, and satellite television, new provisions governing these communications have been added to the Act.

Conductor: Matter that transfers electrical and radio waves.

Coordinated Universal Time: Previously known as Greenwich Mean time.

Counter: *See Frequency Counter.*

Critical Angle: The angle at which radio signals refract in the Earth's ionosphere. Measured in reference to the Earth's surface. Lower angles allow for greater distance of travel by way of ionospheric refraction.

CTCSS: Continuous Tone-Coded Squelch System (CTCSS) also known as tone squelch, is designed to allow users on a shared frequency to hear only users in their user group. Each group is assigned its own squelch tone, and the radio only plays to the audio when the squelch tone is transmitted. The transmitter adds a unique sub-audible code to the transmission

Cutoff Frequency: The frequency where a filter will reject signals.

D

D-layer: Approximately 25-50 miles above Earth's surface, the D-Layer is the lowest region of the Ionosphere. The D-Layer absorbs signals passing through it. The D-Layer disappears soon after sunset, and on short days may not form.

DAC: Digital to Analog Convertor

dB, decibel: A measurement of sound defined by a logarithmic scale between power levels. 3 decibel increase is equivalent to double the power, while 20 dB increase is 100 times the power.

DCS: Digital Coded Squelch. Digital version of CTCSS.

Dead Zone: Area where radio reception of a signal is not possible due to difficulties in propagation.

DECT: Digital Enhanced Cordless Telecommunications is a digital communications standard used by cordless telephones.

Delay: The length of time for which a scanner pauses after a transmission ends before moving on to the next channel.

Dielectric: Matter that insulates radio waves. Generally nonmetallic matter.

Dipole: Antenna made up of two wires connected in a straight line. One wire connects to the radio and other wire connects to group.Usually made to operate best on a specific frequency. Commonly used as a base antenna.

Discone: An omnidirectional, vertically polarized antenna with a disc shape.Discone antennas are broadband antennas (able to receive a wide range of frequencies) that are generally base-mounted.

Discriminator Output: Provides access to the unfiltered audio signal from a radio. This is often required to decode data signals over 1200 Baud.

Direct Communication: Communication between stations without using a repeater.

Direct Wave: Radio signal propagated by line of sight.

Drift: Gradual frequency change in a transmitter or receiver. Commonly occurs when the temperature of a radio changes.

Duplex: Transmission and reception occur on separate frequencies, functions like a telephone, allowing simultaneous talk on both ends.

Dynamic Range: A description of a receiver's ability to receive strong signals without being over-loaded.

E

E-Layer: 50 to 90 miles above the Earth, the E-Layer is a part of the Ionosphere. The E layer absorbs energy from signals that pass through it. Several hours after the sun sets, the E-Layer disappears. This is why AM broadcast stations must lower transmission power at night.

Effective Radiated Power, ERP: Output of a transmitter multiplied by the gain of the attached antenna.

F

F-layer: Responsible for most long distance propagation below 30 MHz, the Ionosphere is the region of the Earth's atmosphere approximately 90 to 400 miles above the surface. Solar heating can cause the F-Layer to split into the F1 and F2 Layers.

Feedline: Cable that connects the radio to the antenna.

Filter: Allows certain frequencies to pass and rejects other frequencies. May be a device or a circuit.

Flutter: Commonly caused by variations in propagations, flutter is the rapid variation in a station's signal strength.

Frequency Modulation, FM: Modulation which varies the carrier frequency of the transmitter according to strength variations in the modulating signal.

Frequency: When used in the context of radio, is the number of times that a wave occurs in a set amount of time. Measured in hertz.

Frequency Counter: A device used for measuring the frequency of a radio wave. Can be used to identify the frequency used by a transmitter.

Frequency Step: The intervals of frequency that the tuner changes when adjusted.

G

Gain: When an antenna appears to increase the signal transmitted or received.

GHz, gigahertz: Unit of measurement for radio frequency waves equal 1000 megahertz or 1,000,000 kilohertz or 1,000,000,000 Hertz

Ground: A wired connection, usually to the Earth, to a zero voltage point.

Ground wave: A radio wave which propagates along the Earth's surface.

H

Harmonic: Frequencies that are multiple of a lower frequency.

HF, High Frequencies: 3 to 30 MHz. Note: HF is commonly used to refer to 1.7-30 MHz

High Pass Filter: A high pass filter allows frequencies above a specified point through, and rejects those below that point. See also Low Pass Filter.

Horizontal Polarization: Antenna that receives or transmits best radio waves with an electrical field parallel to Earth's surface.

Hz, Hertz: Frequency is measured in Hertz. Hertz denotes the number of cycles per second in radio frequency, with one Hertz denoting one cycle passing per second.

I

Image: False signal when produced by circuitry within the receiver.

Indirect FM: *See phase modulation.*

Input Frequency: When used in the context of repeaters, describes the frequency the repeater listens to and retransmits signals received on this frequency.

Intermodulation, Intermod: Spurious or false signals. Occur when multiple signals mix in a receiver or repeater.

Ionosphere: The region approximately 40 to 400 miles above the Earth's surface. This electrically charged region refracts radio signals.

K

Key Up: Slang term for pressing the push to talk (PTT) switch on the radio so items said into the microphone will be transmitted.

kHz, Kilohertz: 1000 Hertz

L

Law of Inverse Squares: The signal strength of a radio wave is inversely proportional to the square of the distance from the source. Because of signal strength does no grow in a linear fashion the closer you get to a transmitter. Instead signal strength will dramatically increase the closer you get to a transmitter.

Line of Sight: As the name implies, communication between two stations that are within sight of each other.

Lockout: A scanner feature that allows for the exclusion of specified channels from the scanning sequence. Can also be applied to the search feature on a scanner to exclude specific frequencies from the search sequence.

Low Pass Filter: Filter of circuit that allows frequencies below a defined point through, and rejects frequencies higher than the defined point. See also *High Pass Filter*.

LSB, Lower Sideband: Lower frequency sideband that transmitter carrier. See Single Sideband Modulation.

Low Threshold: How weak a signal the radio is able to pick up. Used to determine the sensitivity of a radio.

Lowest Usable Frequency, LUF: The lowest frequency that propagate between two points.

LTR: Logic Trunked Radio, or LTR, systems use a transmission protocol developed by the E.F. Johnson company. LTR is used primarily in single site applications.

M

Maximum Usable Frequency, MUF: The highest frequency that can propagate between points.

Medium Wave: Often used to describe any AM signal (540 hHz to 1700 kHz) Medium wave is technically signals between 300 to 3000 kHz.

MHz, Megahertz: 1,000,000 hertz, 1000 kilohertz.

Modulation: Altering the output carrier of a transmitter to encode information into a radio wave. Frequency Modulation (FM) is the most popular types of analog modulation used over 30 MHz.

Monoband Antenna: Antenna designed to operate on only one frequency band.

Multiband Antenna: Antenna designed to operate on multiple frequency bands.

Multihop: Describes a signal that is refracted multiple times between transmission and reception.

Multipath: occurs when the signal from a transmitter is bounced around enough that it arrives at the receiver at different times; the signals arrive at different points in the phase. Also known as Ghosting.

N

Notch Filter: Combines high and low pass filters and only allows a small section of the spectrum through.

O

Omnidirectional Antenna: Describes an antenna can transmit and receive in all directions equally.

Output Frequency: In the context of repeaters, the frequency on which the repeater will retransmit received signals on.

Overloading: Interference causing false signals. Caused by strong signals within the frequency range.

P

Passband Tuning: A receiver feature which allows the user to adjust bandpass to get the best reception based on interference.

Path: Route a signal takes from the transmitter to the receiver.

Phase Locked Loop (PLL): Circuit which generates a large range of frequencies in discreet intervals.

Phase Modulation (PM): Much like FM, PM varies transmission carrier frequency in proportion to strength and frequency of the modulating signal.

Phonetic Alphabet: Words used to represent letters of the alphabet to prevent confusion/misunderstanding.

Polarization: Describes whether an antenna transmits or receives best in a vertical or horizontal plane.

POTS: Plain Old Telephone Service. Voice grade telephone service.

Preamplifier, Preamp: Receiving circuit, which can amplify weak signal. Note that a preamp will often create background noise and may cause signal distortion.

Priority Channel: In scanners, the priority channel is a user defined channel to which the scanner will check for activity more frequently or immediately switch if a signal is present.

Propagation: Describes the process of a radio signal traveling from transmitter to receiver.

Q

Quad: A type of directional antenna made up of two wire squares, each one wavelength, a quarter wavelength apart.

R

Repeater: a device that receives a signal and retransmits it, usually at a higher strength, helping radios cover a larger geographical area.

Resonance: when an electric signal can travel from one end of a wire and back in the period of one cycle of the RF frequency

Resonant Frequency: The frequency where an antenna best receives or radiates.

RF Gain: Control that continuously varies receiver sensitivity.

Rubber Duck (Rubber Ducky): A type of antenna, the rubber duck is an electrically short monopole antenna which is sealed in a rubber housing. There is debate as to where the name rubber duck originated.

S

S-Meter: A graph that shows the relative strength of a signal at reception.

Scanner: A radio receiver that switches between multiple channels or frequencies, stopping when there is activity on the channel or frequency.

Scatter: When a signal scatters directly from the ionosphere.

Search: A feature of scanners and some other receivers that scans a frequency at set intervals and stops when a signal is present. Useful for finding transmitters in a set frequency range.

Selectivity: The ability of the unit to pull a signal out of a noisy environment, or in other words, the ability to select between two signals that are close together on the spectrum. Selectivity depends primarily on the phase noise of the synthesizer and associated circuitry

Sensitivity: Describes how well a receiver can pick up weak signals, measured in microvolts.

Shortwave: Commonly used to describe frequencies between 1.7 to 30 MHz, shortwave is technically 3 to 30 MHz.

Single Sideband Modulation, SSB: An analog modulation types popular below 30 MHz. When an AM signal is generate it has twice the bandwidth of the input (baseband) signals. SSB cuts this AM signal in half by removing one of the side bands and transmits only one side of the signal, because of the way AM signals are generated the entire signal can be reproduced with only one side band. Depending on the side band transmitted it can either be an USB (Upper Sideband) and LSB (Lower Sideband).

Simulcast: In a multi-site trunked system all traffic is repeated across all sites in a trunked system regardless of which talk groups are using the site at the time.

Simplex: Transmission and reception occur on the same frequency.

Skip: Describes sky wave propagation that occurs by Ionospheric refraction. See *Sky Wave*.

Sky Wave: Describes when radio waves are refracted by the ionosphere. This is what allows AM stations to be heard far away.

SMA: A type of standard antenna connector used for coaxial connections.

SmartZone: In a multi-site trunked system traffic to specific talk groups are only broadcast by site where that talk group is present.

Spectrum, Radio: Electromagnetic spectrum from about 3 kilohertz to about 300,000 megahertz

Spectrum Analyzer: A device that analyzes the frequency response, distortion and noise of radio frequencies.

Spread Spectrum: Modulation types that incorporate methods that spread a signal over a chunk of the spectrum to make better use of the bandwidth and avoid interference.

Spurious: False signals that occur by over amplifying a signal, overloading part of the radio or are generated by the internal circuits of the radio. Birdies are examples of spurious signals generated by the internal circuits in a radio.

Squelch: A feature in scanners and other radios that silences the receiver until received signal strength exceeds a predetermined threshold.

SSB: Single Side Band Modulation.

Super High Frequencies, SHF: Frequency range over 3000 MHz.

Surface Wave: *See Ground Wave.*

T

Terminal Node Controller, TNC: TNC converts digital signals from a computer into analog signals for radio transmission, and vice versa.

Tone Access: Activation of a repeater with a predetermined tone or tone sequence. The tone must be received before the repeater relays any transmissions.

Trap Dipole: A type of dipole antenna which has multiple coils, also known as traps, that allow the antenna to be effective on multiple bands.

Tropospheric Ducting: occurs when cold and warm air streams meet about 2 kilometers, or approximately 1.25 miles above the Earth. This phenomenon, which is often see during the summer and usually lasts about an hour at a time, creates a "pipe" that allows signals to travel great distances.

Trunking: Radio system using several channels or frequencies, and allows those channels to be shared by a large number of users, in multiple talkgroups, without their conversations interfering with each other. Trunked systems use a control channel, called the trunk, which transmits data packets which allow a talkgroup to carry on a conversation by telling members of a talk group which frequency to communicate on when they key up.

Tuner: Part of a receiver where the signal from the antenna is selected at a specific frequency from other radio wave.

U

UHF: Ultra High Frequency. UHF is the frequency range from 300 MHz and 3 GHz (3,000 MHz)

UHF Low: 450-470 MHz

UHF-T: 470-512 MHz

Unity Gain: Antenna with effective radiated power equal to the transmitter power. Does not have gain or loss

USB, Upper sideband: Higher frequency sideband that transmitter carrier. See Single Sideband Modulation.

UTC, Universal Coordinated Time: *See Zulu Time.*

V

Vertical Polarization: Antenna that best receives radio waves with an electric field that is perpendicular to the Earth's surface.

VHF: Very High Frequency. VHF is the frequency range from 30-300 MHz.

VHF High band: The range of VHF from 150 to 175 MHz.

VHF low band: The range of VHF from 30 to 50 MHz.

VOX: A VOX circuit powers a transmitter on and off automatically when it detects someone speaking into it.

W

Wavelength: The wavelength of a frequency is the distance over which the shape repeats.

Y

Yagi: A Yagi antenna, short for Yagi-Uda, is type of directional antenna.

Z

ZigBee: A low-power consumption protocol for wireless control and monitoring of devices such as light switches and HVAC. ZigBee is becoming more prevalent, especially in green buildings.

Zulu Time: Zulu, which is Z in the phonetic alphabet, is used to describe Coordinated Universal Time, or UTC.

Index

Note: Page numbers followed by *"f"* and *"t"* indicate figure and table respectively.

161

Printed and bound by CPI Group (UK) Ltd, Croydon, CR0 4YY

03/10/2024

01040342-0001